S0-BVO-756

Higher Energy Polarized Proton Beams

(Ann Arbor, 1977)

AIP Conference Proceedings
Series Editor: Hugh C. Wolfe
No. 42

Higher Energy Polarized Proton Beams

(Ann Arbor, 1977)

Editors
A.D. Krisch
A.J. Salthouse
University of Michigan

American Institute of Physics
New York
1978

Copying Fees: The code at the bottom of the first page of each article in this volume gives the fee for each copy of the article made beyond the free copying permitted under the 1978 US Copyright Law. (See also the statement following "Copyright" below). This fee can be paid to the American Institute of Physics through the Copyright Clearance Center, Box 765, Schenectady, N.Y. 12301.

L.C. Catalog Card No. 78-55682
ISBN 0-88318-141-X
DOE CONF- 7710115

TABLE OF CONTENTS

WORKSHOP ON POSSIBILITIES FOR HIGHER ENERGY POLARIZED PROTON BEAMS

October 18-27, 1977 Ann Arbor

Meeting Rooms on the 3rd Floor of the University of Michigan League

Working Subgroups:

Acc. -- Depolarization during Acceleration or Storage of Polarized Protons or Deuterons - Chair. E.D. Courant, CoChair. L.G. Ratner
Exp. -- Experiments on Spin Dependence at Very High Energy - Chair. O. Chamberlain, CoChair. M.L. Marshak
Hyp. -- Polarized Proton Beams Produced by Hyperon Decay - Chair. G. Fidecaro
Pol. -- Polarimeters for Protons or Deuterons - Chair. J.B. Roberts
Sour.-- Polarized Ion Sources and Low Energy Storage and/or Cooling - Chair. H.M. Glavish
Theo.-- Theory of Spin Dependence at Very High Energy - Chair. F.E. Low, CoChair. G.H. Thomas

The Scientific Secretary A.J. Salthouse will meet with each Chairman and CoChairman for 15 minutes each day to prepare a short summary of the day's progress for distribution the next morning

TUES OCT. 18

9:00 Registration and Coffee 9:45 Welcome: ADK 10:00-11:00 Rev. of Theory:GHT Coffee 11:30-12:45 Rev. of Accelerators:EDC
1:00-2:30 Lunch
2:30-4:00 Rev. of Sources HMG Coffee 4:30-5:30 Rev. of Hyperons GF

WED OCT. 19

9:15-10:30 Rev. of Polarimeter:JBR Coffee 11:00-12:30 Rev. of Expts. MLM			
12:30-2:00 Lunch			
2:00 3:30 Break 4:00 5:30	Sour.	Pol.	Theo.

THURS OCT 20

9:15 10:30 Coffee 11:00 12:00	Acc.	Exp.	Sour.
12:30-2:00 Lunch			
2:00 3:30	Hyp.	Theo.	Pol.
Break			
4:00 5:30	Joint Session		

FRI OCT. 21

9:15 10:30 Coffee 11:00 12:00	Acc.	Theo.
12:30-2:00 Lunch		
2:00 3:30 Break 4:00 5:30	Sour.	Exp.

MON OCT. 24

9:15 10:30 Coffee 11:00 12:30	Acc.	Exp.	Sour.
12:30-2:00 Lunch			
2:00 3:30	Theo	Joint Acc. Pol.	
Break			
4:00 5:30		Acc.	Pol.
8:00	Notation Committee		

TUES OCT. 25

9:15 10:30	Hyp.	Sour.	Pol.
Coffee			
11:00 12:30	Joint Session		
12:30 -2:00 Lunch			
2:00 3:30 Break 4:00 5:30	Acc.	Exp.	Theo.

WED OCT. 26

9:15 10:30 Coffee 11:00 12:15	Sour.	Pol.	Exp.
12:15-1:45 Lunch			
1:45-3:15 Sum.of Accel. EDC 3:15-4:00 Sum.of Hyperons GF			

THURS OCT. 27

9:15-10:30 Summary of Sources HMG Coffee 11:00-12:15 Summary of Expts. OC
12:15-1:45 Lunch
1:45-2:30 Sum. of Polarimeters JBR 2:30-3:15 Sum. of Theory FEL Coffee 3:45-4:15 Sum. of Theory FEL 4:30-5:00 Sum of Workshop ADK

INTRODUCTION

This workshop was held in Ann Arbor from October 18 to 27, 1977 to study the possibility of producing a polarized proton beam or storage ring facility at an energy above the 12 GeV presently available at the Argonne ZGS. The majority of the workshop was spent in 6 small working subgroups. The first and last days were spent respectively on reviews and summaries of each subgroup's area by the chairman or co-chairman of each working subgroup. These subgroups were:

1. Polarized Ion Sources and Low Energy Storage
   Chairman: H.M. Glavish, ANAC of New Zealand
2. Depolarization During Acceleration and/or Storage of Polarized Protons or Deuterons
   Chairman: E.D. Courant, Brookhaven National Laboratory
   Co-Chairman: L.G. Ratner, Argonne National Laboratory
3. Polarimeters for Protons or Deuterons
   Chairman: J.B. Roberts, Rice University
4. Polarized Proton Beams Produced by Hyperon Decay at Very High Energy
   Chairman: G. Fidecaro, CERN
5. Experiments on Spin Dependence at Very High Energy
   Chairman: O. Chamberlain, Univ. of California Berkeley
   Co-Chairman: M.L. Marshak, Univ. of Minnesota
6. Theory of Spin Dependence at Very High Energy
   Chairman: F.E. Low, MIT
   Co-Chairman: G.H. Thomas, Argonne National Laboratory

Several possibilities were considered for higher energy polarized proton facilities. In order of increasing proton equivalent lab energy they were:

Polarized deuterons at KEK	6 GeV
Polarized protons at KEK	12 GeV
Polarized deuterons at AGS or PS	~15 GeV
Polarized protons at AGS or PS	~30 GeV
Polarized deuterons at Serpukhov	~35 GeV
Polarized protons at Serpukhov	~70 GeV
Polarized deuterons at SPS or Fermilab	~200 GeV
Polarized protons at SPS or Fermilab	~400 GeV
Polarized deuterons at ISR	~500 GeV
Polarized protons at ISR	~2,000 GeV
Polarized deuterons at Isabelle	~25,000 GeV
Polarized protons at Isabelle	~100,000 GeV

Initially it was not at all clear if any of these possibilities would be feasible. In fact the technical difficulty in passing the various depolarizing resonances almost certainly increases with energy. Because of their smaller g-factor deuterons may be much easier to accelerate without serious depolarization. However this must be weighed against the deuteron's smaller luminosity and the problems with stripping deuterons or else making the Fermi momentum smear and Glauber corrections if they are not stripped before the nucleon-nucleon scattering occurs.

We normally had 3 of the 6 working subgroups meeting simultaneously to encourage everyone to attend some meetings outside his own special area. In fact this produced some cross fertilization between the different groups which made the planning more realistic, and produced some new ideas.

These proceedings contain 4 sections. The first section contains the overall summary, the schedule, and the participant list. It is intended as a brief introduction and summary for non-experts.

The third section contains 3 reviews of the status of various areas prior to the workshop and 7 short "outline previews" that were distributed to participants about a month before the workshop. These reviews and previews introduce and define some of the items discussed in the summaries. The Appendix contains various items such as the Ann Arbor Convention on Notation and several talks that workshop participants were asked to give and write up. Also included are the working notes on each subgroup session. These were prepared each day by the scientific secretary, A.J. Salthouse, for distribution to all participants. They were not reviewed in detail by each chairman. We include them to insure that no topic is overlooked, but they should be read with the understanding that they were not approved for scientific accuracy by the chairmen or the workshop participants.

The second section contains the summaries of each of the working subgroups, and is the core of the proceedings. These summaries were initially written by each subgroup chairman, but were submitted to all workshop participants for their comments, criticisms, and suggested changes and additions. The editors and the chairmen made a considerable effort to incorporate the many comments and criticisms into the summaries. To the extent that we succeeded, the 6 summaries represent the average opinion of the 37 members of the workshop on the present status of each of the 6 areas necessary for the development of a higher energy polarized proton beam.

We would like to thank Argonne Universities Association, the Department of Energy, and the University of Michigan for providing financial support for this workshop.

ADK and AJS, Ann Arbor, March 1978

POLARIZED PROTON WORKSHOP PARTICIPANTS

	Institution	Dates
Ashkin, J.	Carnegie-Mellon	17-27
Chamberlain, O.	Berkeley	15-27
Chamberlin, E.P.	Los Alamos	17-27
Cho, Y.	Argonne	17-27
Cool, R.L.	Rockefeller	25-27
Cork, B.	Berkeley	17-27
Courant, E.D.	Brookhaven	17-26
Crosbie, E.A.	Argonne	17-27
Derrick, M.	Argonne	26-27
Dick, L.	CERN	17-27
Diebold, R.E.	Argonne	17-27
Fidecaro, G.	CERN	17-27
Glavish, H.F.	ANAC	17-27
Jones, L.W.	Michigan	17-27
Koehler, P.	FNAL	23-27
Krisch, A.D.	Michigan	17-27
Kubischta, W.	CERN	17-27
Leader, E.	Westfield-London	17-27
Loeffler, F.J.	Purdue	24-27
Low, F.E.	MIT	17-27
Madansky, L.	Johns Hopkins	23-27
Margolis, B.	McGill	18-24
Marshak, M.L.	Minnesota	17-26
Michel, L.	Bures-sur-Yvette	17-27
Montague, B.W.	CERN	17-27
Neal, H.A.	Indiana	26-27
Overseth. O.E.	Michigan	17-27
Parker, E.F.	Argonne	19-27
Peaslee, D.	Dept. of Energy	18,26-27
Poirier, J.	NSF	27
Ratner, L.G.	Argonne	17-27
Roberts, J.B.	Rice	17-27
Soffer, J.	Marsailles	17-27
Terwilliger, K.M.	Michigan	17-27
Thomas, G.H.	Argonne	17-27
Turrin, A.	Frascati	17-27
Yokosawa, A.	Argonne	17-27

Scientific Secretary

Salthouse, A.J. Michigan

Secretaries

Anne Allen
Alice Carroll

OVERALL SUMMARY

A.D. Krisch
Randall Laboratory of Physics
The University of Michigan
Ann Arbor, Michigan 48109 USA

Let me begin by saying that I am extremely pleased by the results of this workshop. The 38 experimenters, theorists, and accelerator physicists from 23 institutions in Europe and North America have all worked very hard for 10 days. This workshop concentrated on planning for a type of high energy physics, rather than planning for a specific accelerator project at a specific laboratory. We studied polarized proton and deuteron beams and storage rings in considerable detail for the AGS, PS, SPS, Fermilab, ISR and ISABELLE and in some detail for KEK and Serpukhov. Our goal was to seek a feasible way to study spin-spin forces in strong interactions at very high energy. The workshop was more successful in attaining this goal than we had even hoped would be possible.

In 1/2 hour I could not possibly cover all the important new ideas that were developed during the past 10 days. Fortunately the chairmen of the six working groups have already done an excellent job of summarizing the significant contributions of each working group. I will instead just try to list a few topics that I feel need additional work, and list a few of the major highlights which produced the feeling of optimism, that I think we all share.

One extra activity at the workshop was to establish an international organizing committee to insure the continuation of the Symposia on High Energy Physics with Polarized Beams and Polarized Targets. These symposia were held at Argonne in 1974 and 1976 under

ISSN: 0094-243X/78/007/$1.50

the sponsorship of the ZGS users group. The increased activity in spin studies makes a broader sponsorship now seem appropriate. The IIIrd Symposium will most likely be held at Argonne in October 1978, and the 1980 Symposium will probably be held in Europe. We will soon fix the exact dates, and details. We hope to increase further the participation by people studying spin effects in electron scattering. We also hope to see many of you attending.

A second activity was the effort led by the notation committee under Elliot Leader's chairmanship. Hoping to reduce the confusion caused by different groups using different symbols for the same spin parameter, a major effort was made to agree on a uniform notation. This effort was clearly very painful to many of us, and took much more time and work than I expected. However almost all of us did finally agree on a convention which is published in these proceedings. I hope those few who are not totally happy with every detail of the convention will nevertheless use it. I believe it is ultimately in everyone's best interest to use uniform notation.

PROBLEMS NEEDING EXTRA WORK

I will briefly mention a list of problems that I feel we did not cover in enough detail.

1. Radiation Resistant Polarized Targets

Since spin seems especially important at high-$P_{\perp}^2$ where cross sections are small, high intensity beams are required. At high intensities Polarized Proton Targets suffer radiation damage and will no longer properly polarize. Some studies have been made of "annealing," and there have been searches for radiation resistant materials that are polarizable. More work is clearly needed here and perhaps the polarized target experts should have a workshop on this subject. Perhaps we erred in not inviting them to this workshop.

2. "Low Junk" Polarized Targets

Since the theorists feel that inclusive spin-spin interactions are especially important, one needs "low-junk" polarized proton targets. The "junk" atoms of carbon and oxygen in present targets are very good at producing inclusive pions and protons, which must be subtracted and make precise experiments very difficult. In fact, so far there have been no measurements of inclusive spin-spin forces. Of course, polarized storage rings would totally eliminate this problem, but they do not yet exist. Thus we should again urge the polarized target experts to search for new polarizable materials.

3. Internal Polarimeters

Jumping or avoiding depolarizing resonances will clearly be much more difficult in strong focusing accelerators and storage rings. Thus internal polarimeters will be very important, for they allow resonances to be studied without extracting the beam to an external polarimeter. A gas jet internal polarimeter could simultaneously measure the polarization above and below each resonance. While both the experimenters and accelerator experts did considerable work on this problem, I feel that even more work is needed because of the many interfacing problems.

4. Spin Flip in Storage Rings

The pulse by pulse flipping of the spin at the ZGS is absolutly vital for eliminating systematic errors in high precision experiments. The experimenters made some effort to communicate this to the accelerator people; but I want to stress the need for flipping the spin in storage rings at some regular interval (typically 1 sec to 1000 sec). The accelerator people did some work on this but I do not believe a clearly workable idea emerged.

5. Relation of Large-$P_\perp^2$ Spin Effects to Quarks

There was a general feeling that the large spin effects seen in high-$P_\perp^2$ p-p interactions are caused by the direct interactions of the spin-1/2 quarks, if they exist. But most of the calculations seemed somewhat model dependent. One should try to search for general relations that only assume that quarks have spin 1/2.

6. Polarized Deuteron Acceleration

One should study more carefully the acceleration of polarized deuterons, especially the need for a well calibrated deuteron polarimeter. Such studies might be tried during the planned polarized deuteron run at the ZGS in late 1978. If some unexpected problem makes polarized proton acceleration at AG machines impossible we should have a detailed polarized deuteron backup plan.

HIGHLIGHTS

1. High Intensity Source

Clearly the major highlight was the apparent breakthrough by the polarized ion source experts [H.F. Glavish, ANAC; E. Chamberlin, Los Alamos; W. Kubishta, CERN; and E.F. Parker, Argonne]. They produced a simple new idea which they believe will increase the polarized source intensity by a factor of about 30. The scheme uses the ANAC atomic beam type source that already exists at Argonne and CERN, but bombards the polarized neutral hydrogen atoms in the ionizer stage with D^- ions instead of electrons. The cross section for:

$$D^- + \underset{\uparrow}{H^\circ} \rightarrow \underset{\uparrow}{H^-} + D^\circ$$

is two orders of magnitude larger than the cross section for the present process:

$$e^- + \underset{\uparrow}{H^\circ} \rightarrow \underset{\uparrow}{H^+} + e^- + e^-$$

Therefore the source experts expect the polarized ion source intensity to increase from its present 50 → 100μa to perhaps 1 → 5ma. This scheme was carefully studied for a week, and I believe no one could find any flaw in the scheme. We are all eager to see if the source experts can get this improvement working in the next 6 months. Notice that the $H^-_\uparrow$ ions can be injected into accelerators with better efficiency than the present $H^+_\uparrow$ ions.

2. Depolarization in Strong Focusing Accelerators

The accelerator experts, led by E.D. Courant (Brookhaven), pointed out that while depolarizing resonances are certainly worse in strong-focusing accelerators than in the weak-focusing ZGS, they may not be as bad as was once feared. In fact, for polarized deuteron acceleration the resonances are not very strong at all. The ratio G_d/M_d is 25 times smaller than G_p/M_p so the depolarizing deuteron resonances are 25 times further apart than the proton resonances. Thus there is only 1 depolarizing deuteron resonance at the PS or AGS which is fairly weak, and there are about 10 at the SPS or Fermilab. Thus polarized deuterons look fairly easy at 30 GeV or indeed at 300 GeV.

However, the experimenters made it clear that they would much prefer polarized protons. The accelerator experts now decided that accelerating polarized protons in strong focusing accelerators might somehow be possible if a sufficient technical effort is made. In fact Professor Courant calculated that, using ZGS-type resonance-jumping schemes [pulsed quadrupoles for intrinsic resonances and pulsed orbit bumps for imperfection resonances], a polarized proton beam could probably be accelerated to almost 25 GeV at either the AGS or PS without very serious depolarization. Much of the new optimism comes from the ZGS success in repeatedly jumping 29 depolarizing resonances with no significant depolarization up to 12 GeV. This gives everyone confidence that the Froissart-Stora equations adequately describe depolarization during acceleration in synchrotrons and can be used to precisely calculate the necessary corrections. Courant compared acceleration to 25 GeV at the AGS (which is rather similar to the PS) with acceleration to 12 GeV at the ZGS. The AGS has about twice as many imperfection resonances and they are each about 10 times stronger than the ZGS imperfection resonances which each typically cause 5% depolarization.

Because the basic AGS periodicity is 12 while the basic ZGS periodicity is 4, there are only 6 or 8 intrinsic AGS resonances compared with the 10 intrinsic ZGS resonances. The AGS intrinsic resonances are estimated to be typically 3 times stronger than at the ZGS, where many are already strong enough to each totally depolarize the beam. A carefully planned program of designing upgraded ZGS-type correction magnets would probably allow polarized protons to be accelerated to almost 25 GeV without serious depolarization. This would probably require additional studies of depolarizing resonances at the ZGS polarized beam, but there seems to be no fundamental problem that some money and a lot of effort and thought could not solve. Some very strong depolarizing resonances (intrinsic and imperfection) occur near 25 GeV in the AGS and near 22 GeV in the PS. It is not clear if they could be passed without great effort.

3. Spin-Spin Forces at High-$P_{\perp}^2$

Two general conclusions about spin and high energy physics were more or less agreed upon by the experimenters led by O. Chamberlain and the theorists led by F.E. Low:

A. To understand fundamental interactions one needs precise spin experiments with both electron and proton accelerators

i) e-p scattering gives information about the "quark" wave function.

ii) p-p scattering gives information about the "quark-quark" interaction.

B. The spin-spin forces in very high energy large-$P_{\perp}^2$ proton proton interactions may be a key to understanding the "quark-quark" force. The spin-spin inclusive experiments may be even more important than the spin-spin elastic experiments.

4. The Siberian Snake

The "Siberian Snake" scheme for eliminating depolarizing resonances recently proposed by Derbenev, Kondratenko and Skrinsky (Novosibirsk), has the protons polarized in the accelerator plane rather than vertically. A solenoid magnet is placed in one straight section and tuned until the polarization vector returns to the same orientation after each pass around the synchrotron. This scheme is simple and elegant in theory. If it works, it could totally eliminate depolarization, even up to full energy, at Fermilab, SPS, or ISABELLE. But it clearly needs much more effort and thought.

EVENT RATES

I will finally discuss the event rates that can be expected for various measurements of spin effects at high $P_\perp^2$. I calculated the rates for elastic and inclusive events at $P_\perp^2 = 3(\text{GeV/c})^2$ where I expect spin-spin effects will be large even at very high energy. For elastic events I took

$$\sigma = \frac{d\sigma}{dt}\,\Delta t = [10^{-32}\ \frac{\text{cm}^2}{(\text{GeV/c})^2}][10^{-1}(\text{GeV/c})^2] = 10^{-33}\text{cm}^2$$

For inclusive events I took

$$\sigma = E\frac{d^3\sigma}{dp^3}\ [P^3\frac{\Delta P}{P}\Delta\Omega] = [2\ 10^{-30}\ \frac{\text{cm}^2}{(\text{GeV/c})^2}][2\ 10^{-2}(\text{GeV/c})^2] = 4\ 10^{-32}\text{cm}^2$$

The number of events per day is calculated from the luminosity, L, using

$$\text{Events/day} = 10^5(\text{sec/day})\,L(\text{Protons}^2/\text{cm}^2\text{-sec})\,\sigma(\text{cm}^2)$$

For fixed target experiments L is given by

$$L = I_o \left(\frac{\text{Beam Protons/burst}}{6\ \text{sec/burst}}\right) 4\ 10^{23} \left(\frac{\text{Target Protons}}{\text{cm}^2}\right)$$

The polarized target was taken to be 10 cm long with an effective density of .07. For the polarized gas jet we assumed there were enough multiple traversals ($\sim 10^5$) to give about 10^{28} cm^{-2}/6 sec burst for a luminosity of 2 10^{27}/ cm^2-sec. The luminosity for polarized colliding beams was taken to be 10% of the present ISR luminosity of 4 10^{31}. Notice that we limited the unpolarized beam intensity to 3 10^{11} because of radiation damage to the polarized target.

In calculating the errors in A and A_{nn} we took P_B=70% and P_T=70%. The error in the analyzing power is given by

$$\Delta A = \frac{1}{P_T \sqrt{\text{events}}} \quad \text{or} \quad \Delta A = \frac{1}{\sqrt{(P_B^2 + P_T^2)\ \text{events}}}$$

depending on whether one or both of the incident protons are polarized. The error in the spin-spin correlation parameter is

$$\Delta A_{nn} = \frac{1}{P_B P_T \sqrt{\text{events}}}$$

For inclusive processes with colliding beams or gas jets these same formulae give the errors. However for inclusive processes with polarized targets, 90% of the events come from the "junk" oxygen and carbon; thus the errors are much larger.

As can be seen from the error column, either polarized colliding beams or an accelerated polarized beam at FNAL or SPS are necessary to measure very high energy spin-spin interactions at $P_\perp^2 = 3\,(\text{GeV}/c)^2$. However, the hyperon decay beam and the polarized gas jet may be important intermediate steps.

ELASTIC AND INCLUSIVE SPIN EXPERIMENTS AT $P_{\perp}^2=3(GeV/c)^2$	s $[GeV^2]$	Luminosity $[cm^2\text{-}sec]^{-1}$	Events/Day Elastic	Events/Day Inclusive	ΔA Elastic	ΔA Inclusive	ΔA_{nn} Elastic	ΔA_{nn} Inclusive
					←	Run of 1 Day		→
Unpolarized beam on polarized target [SPS or Fermilab] $3\ 10^{11}$/6 sec burst	~800	$2\ 10^{34}$	$2\ 10^6$	$8\ 10^7$	0.1%	0.2%	Not Possible	
Unpolarized beam on polarized gas jet [SPS or Fermilab] $3\ 10^{13}$/6 sec burst	~800	$2\ 10^{27}$	.2	8	320%	50%	Not Possible	
Polarized protons from hyperon decay on polarized target [SPS or Fermilab] $1.5\ 10^6$/6 sec burst	~500	10^{29}	10	400	32%	70%	65%	100%
Polarized Colliding beams [ISR or ISABELLE] 10% of Present ISR Luminosity	~2000	$4\ 10^{30}$	400	16,000	5%	1%	10%	1.6%
Polarized beam on polarized target [SPS or Fermilab] $3\ 10^{11}$/ 6 sec burst	~500	$2\ 10^{34}$	$2\ 10^6$	$8\ 10^7$	0.07%	0.2%	0.14%	0.2%

CONCLUSION

Perhaps I can best summarize the tone of the workshop by stressing the optimism produced by the source experts and the accelerator experts. If ZGS-type corrections for depolarizing resonances work at the AGS or PS, then one should be able to reach almost 25 GeV without serious depolarization, inject into ISABELLE or ISR, and operate near $s = 2000\ GeV^2$. If the new ion source scheme really gives a factor of 30 gain in intensity, the colliding polarized beam luminosity will be increased 1000-fold to within a factor of 10 of the present ISR luminosity. While both the above sentences start with if, there was a strong feeling that in the 1980's we might be studying p-p inclusive cross sections in pure spin states in clean colliding beam experiments.

I want to conclude by thanking all you distinguished scientists for 10 days of very hard work. I think we are all exhausted; but I also think that the possibility of studying spin-spin forces at very high energy makes it seem worthwhile.

INDIVIDUAL SUMMARIES

EXPERIMENTAL SESSIONS--ANN ARBOR WORKSHOP

Owen Chamberlain-Chairman-Berkeley
M.L. Marshak-Cochairman-Minnesota

This report outlines the thoughts of the workshop participants concerning high energy spin dependence experiments of the future.

Attempting to look ahead as best we can, we have concentrated on experiments thought to give information on "quark-quark" scattering or more precisely on the direct scattering of the proton's constituents. There was more emphasis on inelastic processes, particularly inclusive experiments at high momentum transfer than on elastic scattering. We are presuming that the strong theoretical interest in inclusive processes at high transverse momentum will continue even if there are changes in the basic theory. Needless to say, our present suggestions will be couched in terms of present day theoretical ideas.

We believe it is very important to measure and verify those aspects of any theory that involves the spin of the interacting particles. Sometimes the spin effects may be inferred from general symmetry principles. However, in many cases it may be possible to check the spin dependence of the forces directly predicted by a particular theory. Fairly extensive polarization experiments therefore seem obviously to be warranted, provided they are feasible.

While few would doubt that strong parallel efforts are needed both with hadron beams and lepton beams, we point out that at the present stage of the theory this is particularly important. We need lepton and photon interactions to probe the parton or quark structure of

ISSN: 0094-243X/78/020/$1.50

the nucleon. At the same time, we must have hadron beams if we are to study gluons, which are thought not to interact directly with leptons and photons. From a slightly different point of view, we need electromagnetic and weak interactions to determine the "quark" wave function of the nucleon. Then we need hadron-hadron interactions to allow us to study "quark-quark" interactions.

SIZEABLE SPIN EFFECTS ALREADY SEEN

There are already some indications that spin effects are quite large in inclusive processes at high energy. One outstanding example is the experiment of Bunce, *et al.*, on inclusive Λ production from proton-beryllium collisions at Fermilab. The polarizing power of the Λ in the reaction $p+Be \rightarrow \Lambda$+anything is observed to be 30% at $P_\perp$ of 1.5 GeV/c near $x_F=0.6$. (Here x_F is the Feynman x defined as the parallel component of the Λ momentum divided by the maximum c.m. momentum.) We are inclined to the view that these results, by themselves, indicate that the underlying quark-quark interactions are highly spin dependent, though we are aware that this view may be controversial. This interpretation would be more acceptable if $P_\perp$ were larger.

Another indication of strong spin dependence at very high energies is found in the Indiana University results on the polarizing power of protons from the inclusive reaction $p+C \rightarrow p$+anything, where the target is a fine carbon filament that can be moved through the beam of circulating protons in the main ring at Fermilab. At a range of Fermilab energies they find the polarizing power of this inclusive reaction to be about 8%. These are measurements at x_F of about -0.9 and $P_\perp$ about 1 GeV/c.

Of course, large spin effects are observed at 12 GeV/c at the ZGS. In the inelastic process $p_\uparrow+p \rightarrow \Delta^{++}+n$

the analyzing power is -40% at 6 GeV/c at medium $P_\perp$, and appears energy independent. The analyzing power for elastic p-p scattering is as large as 16% at medium $P_\perp$. The spin-spin correlation parameter, A_{NN}, for $p+p \rightarrow p+p$, sharply increases to about 30 percent near $P_\perp = 2$ GeV/c which is the highest momentum transfer studied.

There is no indication of large spin effects in the elastic scattering experiments that have been performed to date at very high energy. The analyzing power, A, in elastic p-p scattering has been measured at 100 GeV and 300 GeV at Fermilab. There are very preliminary results from CERN at 150 GeV. None of these give indications of any large polarization effect being evident. However these experiments are limited to the small $P_\perp$ diffraction peak, and have not yet extended out to the large $P_\perp$ region where the spin-spin force is so large at ZGS energies. Moreover they are limited to one-spin experiments, while two-spin cross sections show the largest spin effects at ZGS energies.

In summary, it seems that at very high energy the spin-orbit (one-spin) effects in p-p elastic scattering are quite small at low momentum transfer. High energy inclusive cross sections, however, show significant spin dependence even at small $P_\perp$. This seems to be in at least rough accord with present-day theory, in which the elementary "quark-quark" interactions are more directly revealed through inelastic processes than elastic, except possibly at large $P_\perp$. The elastic scattering may well show large two-spin and possibly one-spin effects if they can be measured at very high momentum transfer.

POSSIBLE POLARIZED BEAMS ABOVE 12 GeV

Because thorough spin studies call for polarized beams, obviously one must ask if polarized protons can be accelerated to very high energies. At the beginning of the Workshop most participants thought the answer was "No", but in the course of the Workshop we gained the impression that the answer is really "Maybe". Apparently there is now much greater faith in the reliability of the depolarization theory, based particularly on the experience with the ZGS polarized beam. The accelerator section of this Workshop concluded, that it should now be possible to maintain good beam polarization up to 25 GeV in strong focusing accelerators such as the AGS, with only moderately greater difficulty than was overcome at the ZGS. (See the summary of the accelerator section.)

We also learned that it should be remarkably easy to accelerate deuterons without losing polarization, since the deuteron anomalous magnetic moment is quite small. In fact, a given depolarizing resonance occurs for deuterons at an energy about 25 times as great as for protons.

In spite of the ease of obtaining polarized deuteron beams it was generally felt that the polarized proton beams were the most desirable. This feeling was based on the higher nucleon energy obtainable and especially on the cleaner interpretation of experiments. Nevertheless, some effort was devoted to polarized deuteron beams during the workshop.

A recently developed promising idea is a polarized proton beam obtained from Λ decay at very high energy. A formal proposal has been submitted at the CERN SPS by one group, and calculations are in progress for a similar beam at Fermilab. These will not be discussed here in detail because they are covered in another summary. These beams may be the only polarized beams available for some time at SPS or Fermilab energies. They could be very important in opening the way to spin studies at high energy.

JETS

Collision processes that produce particles with high transverse momentum ($P_\perp$) are thought to be mainly attributable to direct "quark-quark" interactions. The same may be true for a high-$P_\perp$ jet of particles since, quark confinement implies that a single quark resulting from a "quark-quark" scattering will frequently manifest itself as a jet of particles moving with the $P_\perp$ of the quark. The jets may be very important in polarization experiments because the differential cross section for producing a jet of a given transverse momentum is much greater than the cross section for producing a single particle of the same transverse momentum. According to these views jet production experiments on polarization at high $P_\perp$ may be possible and very important.

There is presently no direct knowledge of the minimum $P_\perp$ required before a reaction such as jet production starts to directly reveal "quark-quark" scattering. Presumably the knowledge of this minimum $P_\perp$ must come in the future. Fortunately at Fermilab energies a group of particles that appears as a jet in the center-of-mass system also looks like a jet in the laboratory system since the particles involved are highly relativistic in both the c.m. and lab systems. This may make jets easier to identify and define.

According to G.H. Thomas's introductory review talk, the 2-spin measurements, such as A_{NN}, may show much greater spin-dependent effects than 1-spin measurements such as A at very high energy and high $P_\perp$. This is because spin-spin forces in "quark-quark" scattering may be much larger than spin-orbit forces.

SPECIFIC IDEAS FOR EXPERIMENTS

We shall now evaluate certain specific experiments in order to determine what types of beams and targets might be required to study the physics of spin at high energy. In order to focus on experiments which are feasible in the next decade, we first summarize the probable characteristics of existing or proposed polarized beams and targets in tables I and II.

TABLE I. CHARACTERISTICS OF POLARIZED BEAMS

BEAM	DATE	ENERGY (GeV)	INTENSITY (PROTONS/ BURST)	POLARIZATION
Argonne ZGS	1977	12	$4x10^{10}$	70%
p beam from Λ decay (SPS or FNAL)	1979	100-300	10^{7}	40%
d beam (SPS or FNAL)	>1980	100-1000	10^{11}	60%

TABLE II. CHARACTERISTICS OF POLARIZED TARGETS

TARGET	DILUTION FACTOR	PROTONS/ cm^2	POLARIZATION	COMMENT
Ethylene glycol, etc	8	$6x10^{23}$	80%	conventional target
Ammonia	5	$6x10^{23}$	80%	hard to reverse spin
D-H	1 or 3	$6x10^{23}$	32%	not yet operational
Polarized jet	1	10^{12}	90%	high reversal frequency possible

In table III we summarize some possible combinations of polarized beam <u>and</u> target. Obviously there are classes of easier experiments in which only one particle in the initial state is polarized.

TABLE III. EFFECTIVE LUMINOSITY OF CERTAIN POLARIZED BEAM-POLARIZED TARGET COMBINATIONS. OUR DEFINITION IS: EFFECTIVE LUMINOSITY = LUMINOSITY x $(\text{BEAM POLARIZATION})^2$ x $(\text{TARGET POLARIZATION})^2$/(DILUTION FACTOR)

Initial-state particles, both polarized	Effective Luminosity ($cm^{-2}sec^{-1}$)
Argonne polarized beam on conventional polarized target	5×10^{32}
Polarized protons from Λ decay on conventional polarized target (SPS or FNAL)	8×10^{27}
Polarized d beam on conventional polarized p target (SPS or FNAL)	2×10^{32}
Internal polarized d beam on polarized gas jet target	3×10^{26}
Polarized d colliding beams (IRS or ISABELLE)	5×10^{29}

In table III we have used the "effective luminosity," which is the equivalent luminosity of an experiment in which all of the relevant spins are 100 percent polarized and there is no unpolarized material diluting the target. This concept is relatively easy to apply to spin analysis in the initial state. When the spins of final-state particles are studied, there are factors such as the efficiency of decay detection or the analyzing power of a polarimeter which are relevant to any calculation on the feasibility of an experiment. Moreover, the dilution factor should not be used in those "elastic" experiments in which a clean signal can be picked out of the "junk" background. In any event, it is clear that one-spin measurements are easier than two-spin measurements and these are easier than three-spin measurements. Since we seek to study high-$P_\perp$ processes with limited luminosity, we had best measure as few spins as possible.

INCLUSIVE PROCESSES

There are numerous inclusive processes that can be studied in an attempt to learn about the spin dependence of "quark-quark" interactions. Since there currently exists no credible spin-dependent model for such processes there are few criteria for choosing which particular reaction is best for the spin dependence of the "quark-quark" interactions. The most straightforward criterion is ease of measurement since others usually depend on some model about the nature of the scattering process.

The simplest polarization experiments are the one-spin measurements such as the analyzing power, A, or polarizing power, P. In fact, the analyzing power will frequently stand for two quantities, not necessarily equal, depending on whether a polarized beam or a polarized target is used. Frequently neither of these quantities is the same as the polarizing power.

For inclusive reactions the analyzing power is difficult to measure using a conventional polarized target. In an inclusive reaction, no kinematic relationships are available to distinguish between scattering on free and bound protons. Therefore the unpolarized protons bound in nuclei such as carbon, which constitute a large fraction of conventional polarized targets, are responsible for most of the events counted. We describe this difficulty by defining a dilution factor. Consider a target material that is 12.5% hydrogen, 80% polarized; this will act as a target that is 100% nucleons, 10% polarized; and its dilution factor is 8.

The one-spin inclusive cross sections are sensitive to the interference of the single-spin-flip amplitude with the spin-averaged amplitude. Some processes are of obvious interest: $p_{\uparrow}+p \rightarrow p+\text{anything}$ involves meson exchange and should be similar to elastic scattering; $p_{\uparrow}+p \rightarrow \pi+\text{anything}$ and $p_{\uparrow}+p \rightarrow K+\text{anything}$ involve baryon exchange and have been compared to backward meson-nucleon elastic scattering. If current theoretical ideas are correct, the analyzing power, A, for the reaction $p_{\uparrow}+p \rightarrow \text{jet}+\text{anything}$ should be an extremely interesting quantity. (The vertical arrow as a subscript on an initial-state symbol indicates polarized particles; on a final-state symbol it would indicate a particle whose polarization was being measured.)

Other one-spin experiments involve measuring polarizations of final-state particles. These experiments have the advantage of requiring neither a polarized beam nor a polarized target. Experiments studying the inclusive production of hyperons have the advantage that the spin of the final-state particle is analyzed by the decay. This property has already been used in several experiments. Another possibility is to look at final-state particles in the target fragmentation region; these move slowly enough in the laboratory that their spin may be analyzed by a second scattering in a conventional carbon polarimeter. Conceivably, the spin of a quark could be determined by detailed analysis of a jet, but, to our knowledge, there is currently no basis for making such an analysis.

There are also various two-spin measurements that

can be made for inclusive reactions. One set is those experiments in which the spins of both initial particles are known such as $p_{\uparrow}+p_{\uparrow} \rightarrow \pi$+anything. These suffer from the same "dilution" problems as the polarized target analyzing power measurements and so far no such spin-spin inclusive measurements have been made. However they will soon be tried at the ZGS. Measurement of one initial- and one final-state spin are also possible, but, these of course, yield information about a different combination of amplitudes. The experiment $p_{\uparrow}+p \rightarrow \Lambda+$ anything where the Λ spin is analyzed by its decay, is one which could yield accurate data for the depolarization and spin rotation parameters. While these two-spin experiments are more difficult, they may well yield the most precise information about direct "quark-quark" interactions.

Table IV gives estimates of the largest $P_{\perp}$ that could be measured in various one-spin and two-spin inclusive experiments. The calculations assume a polarized proton beam from Λ decay and a conventional polarized proton target, simply because these possibilities may be the first ones available at high energies. Of course, other beam-target combinations could do much better.

If we wish to obtain reasonable statistics on inclusive processes at still higher transverse momenta we must obtain higher luminosities than are available with the polarized proton beam obtained from Λ decay. Such possibilities are included in table III. To decide what combination of polarized beam and polarized target is needed, we ask the question: What luminosity is needed to study the process of inclusive jet production in proton-proton collisions, at $P_{\perp}$ of 4 GeV/c? The Fermilab results indicate an invariant cross section

TABLE IV. LARGEST $P_\perp$ MEASURABLE FOR VARIOUS INCLUSIVE REACTIONS, BASED UPON A 10^7/PULSE POLARIZED PROTON BEAM FROM Λ DECAY AND A CONVENTIONAL POLARIZED TARGET. 5% STATISTICAL ERROR IS ASSUMED.

REACTION	MAXIMUM $P_\perp$ (GeV/c)
One-spin:	
$p_\uparrow + p \rightarrow p$+anything	1.5
$p_\uparrow + p \rightarrow \pi$+anything	1.5
$p_\uparrow + p \rightarrow \Lambda$+anything	2
$p_\uparrow + p \rightarrow$ jet+anything	2.5
Two-spin:	
$p_\uparrow + p_\uparrow \rightarrow p$+anything	1
$p_\uparrow + p \rightarrow p_\uparrow$+anything	<1
$p_\uparrow + p \rightarrow \Lambda_\uparrow$+anything	2
$p_\uparrow + p_\uparrow \rightarrow \pi$+anything	1
$p_\uparrow + p_\uparrow \rightarrow$ jet+anything	2

$Ed^3\sigma/dp^3 = 3 \times 10^{-33}$ $cm^2/(GeV/c)^2$. We choose an interval of 0.1 GeV/c in $P_\perp$, an interval of 0.3 in x_F, and assume, conservatively, that we count jets only in the interval 0.5 radian in azimuthal angle. Then to obtain 400 events in 10^5 seconds we need an effective luminosity value of 2×10^{31} cm^{-2} sec^{-1}. Some of the situations covered in table III provide the needed luminosity, but some do not. One is left with the impression that if the spin dependence of jets is to be studied in full detail then it will be imperative to have direct acceleration of polarized beams in high-energy accelerators such as Fermilab or SPS.

ELASTIC PROCESSES

Although we have argued that inclusive reactions may be a better probe of "quark-quark" interactions than elastic processes, there is no doubt that elastic reactions have a much better developed formalism. For elastic scattering, we know how to relate the spin-dependent observables to the spin-dependent scattering amplitudes. However the direct connection of these elastic amplitudes with the "quark-quark" spin-dependent scattering amplitudes does not yet exist. Despite this theoretical uncertainty there are clearly some intuitively interesting elastic experiments. Measurements at Argonne have shown large two-spin asymmetries in A_{NN} at large momentum transfer in p-p elastic scattering at 12 GeV/c. As mentioned above, preliminary Fermilab data indicate that A is small in p-p elastic scattering up to $t = -1.0\ (\text{GeV/c})^2$. There should be an effort to study high-$P_\perp$ 2-spin elastic experiments at very high energy

The very-small-momentum-transfer elastic scattering is also of interest because of the possibility of measuring the amplitude phases through Coulomb interference. Measurement of the differential cross section in the Coulomb-interference region allows a determination of the real part of the spin-averaged (nuclear) scattering amplitude. Measurement of the analyzing power in this region ($|t|<0.1(\text{GeV/c})^2$) should allow a separate determination of the imaginary part of the single-spin-flip amplitude in the forward direction. Measurements of the total cross section and the spin-dependent total cross section differences allow a full determination of the imaginary parts of the three amplitudes that can be nonzero in the forward direction.

One particular form of elastic scattering that may be especially important is hyperon-proton scattering.

It is already known that Λ's can be produced with large polarization at high energy, and that the Λ reveals its polarization state when it decays. Thus D_{NN} and D_{SS} measurements may well be made for Λ-p scattering before the corresponding measurements are made for p-p scattering.

One final elastic scattering experiment that is both feasible and interesting is Schwinger's proposal for magnetic-moment n-p scattering. For four-momentum transfers of the order of 10^{-4} to 10^{-5} $(GeV/c)^2$, the forward neutron should be almost 100 percent polarized. Demonstration of this high polarization may help in the design of a possible polarized neutron beam.

COMMENTS ABOUT THE BEAMS AND TARGETS REQUIRED FOR THESE EXPERIMENTS

Polarized protons from Λ decay--This appears the least difficult scheme for producing high-energy polarized proton beams. As indicated in table II, its projected intensity of 10^7 protons per burst will permit a number of low-$P_\perp$ experiments, which could increase further the enthusiasm for the acceleration of polarized protons or deuterons. The low flux makes large solid angle detectors imperative, but even a large solid angle will not make this beam usable for high-$P_\perp$ physics.

Polarized gas jet target--The main advantage of the polarized gas jet as a target is the absence of target background from elements heavier than hydrogen. Thus, it has a dilution factor of 1. The polarization is expected to be large and well known. However, the luminosity obtainable with this target is expected to be too low for high-$P_\perp$ physics. In conjunction with a polarized beam, the polarized jet target will permit clean two-spin measurements at small $P_\perp$.

H-D targets--From the data presented here, it is clear that the dilution factor of 8 in conventional polarized proton targets presents considerable experimental difficulties, particularly for studies of inclusive reactions. The deuterium hydride (H-D) target, which has a dilution factor of either one or three, depending on whether the deuteron is polarized, would be of considerable value, even though it has a low maximum polarization. This target would be particularly useful in inclusive reactions where the isospin of the target nucleon is probably not important. In that case, the dilution factor of one would result in considerably reduced statistical and systematic errors. This target must yet be proven operational. One particular problem is that it must be possible to reverse the target spin direction periodically, without significant loss of polarization.

Intersecting polarized beams--It is clear that experiments using different spin orientations are necessary for a full amplitude analysis. For that reason, any polarized intersecting beam facility should have the capability for NN, LL, SS and LS type initial spin states. The NN state is the easiest in a polarized proton storage ring. The other states could be achieved at the interaction region by using precession magnets to rotate the spins into the desired orientation at the crossing region and then return them to the N state after the interaction region. The LS state only requires different precession magnets in each ring. In the case of stored deuterons, precession is almost impossible and consideration should be given to designing a storage ring which stores an L type beam. This may require the "Siberian Snake" scheme described in the accelerator section.

AN OVERVIEW

We have adopted the view that a principal goal of polarization experiments should be the study of "quark-quark" scattering in events with large transverse momentum. Thus we were led to study high-$P_\perp$ processes with very low cross sections and processes in which both colliding particles in the initial state are polarized. There are also many experiments to be done in which only one of the initial-state particles is polarized, and many interesting polarization experiments at smaller $P_\perp$ where the cross sections are larger. In fact, many interesting experiments have been omitted from consideration--some for lack of time and energy, some by oversight.

This summary report contains a number of computational results based on very shaky foundations. In some cases we have estimated the intensity of future polarized beams without having a complete understanding of the required accelerator technology. This report is therefore just a set of suggestions, to which we hope many new ideas will be added. Our conclusions are more to be challenged than believed.

SUMMARY OF THE SUMMARY REPORT

Future theories can only be believed if they correctly explain the spin dependences of scattering processes as well as their differential cross sections. Experiments on spin dependence will therefore be important for the foreseeable future. These polarization experiments seem destined to be expensive and difficult--but they are not impossible and they are necessary.

SUMMARY OF THE THEORY GROUP*

F. E. Low
Massachusetts Institute of Technology, Cambridge, MA 02139

ABSTRACT

This report will be divided into general remarks on three subjects: 1) high $p_\perp$ signatures of the underlying hadronic process, 2) some reminders of general spin dependence properties, and 3) a discussion of polarimetry; followed by a report of some miscellaneous calculations done by members of our group. Some of this work was done by individuals, and some arose out of joint conversations. For this reason, I have not tried to assign credit or blame, except where I have included specific write-ups.

I. HIGH $p_\perp$

In these processes, one may use polarization measurements to seek the signature of the underlying basic process. If, as suspected, this process turns out to be a quark-vector-gluon interaction, then it should (at high energy) have a strong helicity dependence: helicity conservation in $q \to q+V$ (V stands for vector gluon) as well as in $q+\bar{q} \to V$. This principle provides a mnemonic for experiments. Thus in $q+\bar{q} \to V \to s+\bar{s}$, the initial quarks must have opposite helicity, as must the final quarks. The relevant Feynman graph is shown in Fig. 1. If the gluon coupling becomes weak at high energy, then one can, for example, pick out this process by requiring the production of a strange pair by non-strange hadrons, for example, in the process $p+\bar{p} \to$ (a strange high $p_\perp$ jet)+X as shown in Fig. 2, which should show a strong dependence on the initial helicity states, preferring $\lambda_p = -\lambda_{\bar{p}}$, i.e. $S_z = 1$. In an initially unpolarized experiment, one would find a strong helicity correlation between two final strange jets in π+p experiments, as shown in Fig. 3. In these processes, the strange final state is needed to rule out a competing second order process such as shown in Fig. 4.

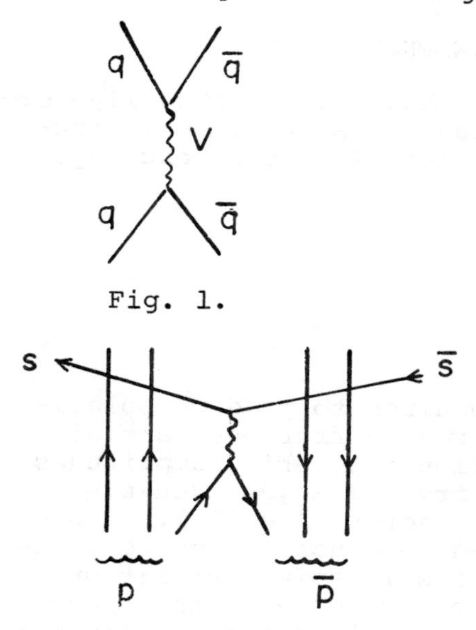

Fig. 1.

Fig. 2.

*Work supported in part through funds provided by ERDA under Contract EY-76-C-02-3069.*000.

ISSN: 0094-243X/78/035/$1.50

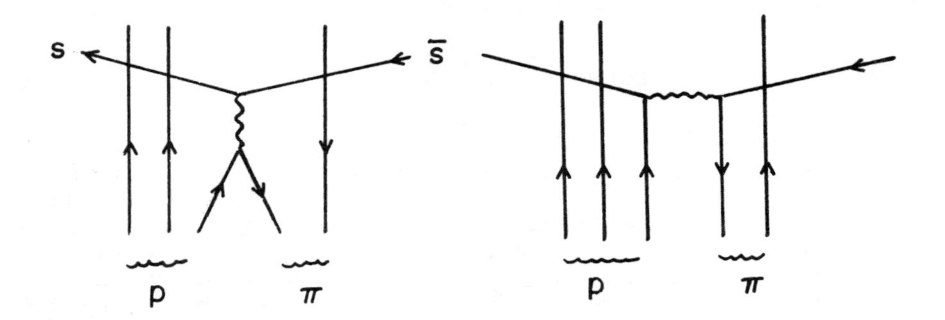

Fig. 3. Fig. 4.

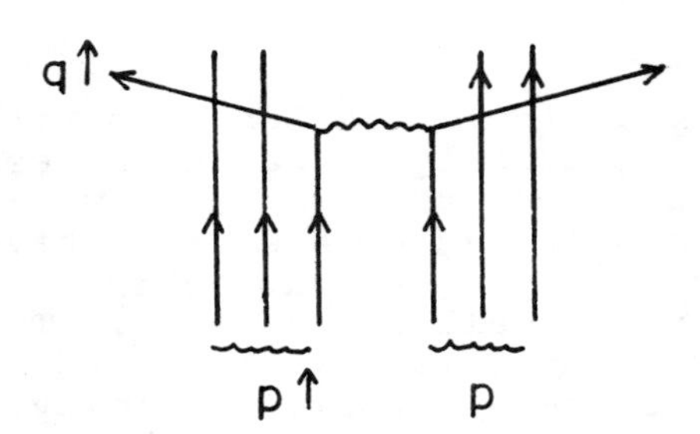

Fig. 5.

On the other hand, typical initial final helicity correlations can be seen in a process like $p{\uparrow}+p\rightarrow$(non-strange jet↑)+X as shown in Fig. 5.

II. SPIN DEPENDENCE

Here I repeat things we all know, but can easily forget. That is, spin dependence can be large and polarization zero; since, for example, in π-N elastic or exchange scattering, with

$$f = f_1 + i\vec{\sigma}\cdot\hat{n}f_2$$

$$P = \frac{2\,\mathrm{Im}\,f_2{}^*f_1}{|f_1|^2+|f_2|^2}$$

so that a phase difference is required to produce polarization. Now at high energies, phase differences are discouraged. For example, diffraction scattering amplitudes appear to be largely pure imaginary and Regge exchange amplitudes have the single phase factor $(1\pm e^{-i\pi\alpha})$. There can nevertheless be strong spin-dependence, which would be measured by a spin rotation, or a spin-spin correlation.

The process $\pi^-p\rightarrow\pi^\circ n$ is a good example of this possibility. *The differential cross-section for $\pi^-p\rightarrow\pi^\circ n$ has a sharp turn-over as $t\rightarrow 0$ (as shown in Fig. 6) at <u>all</u>

*This discussion was given by Leader.

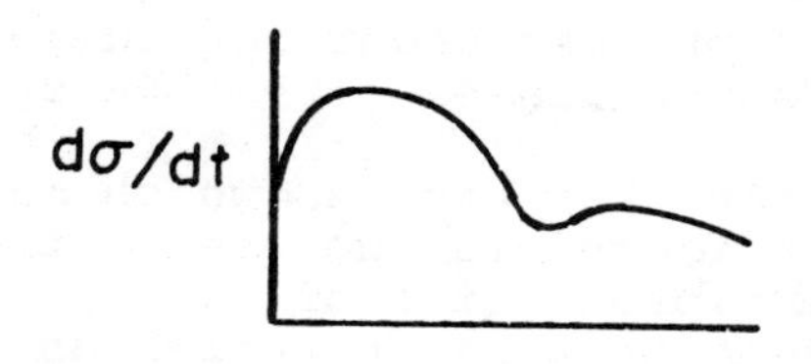

Fig. 6.

energies from 5→FNAL. This is usually explained as due to dominance of the helicity-flip amplitude F_{+-} which <u>must</u> vanish at t=0. Models using ρ exchange find this easy to explain since the ρ likes to flip helicity, as learned from its role in the E.M. form factors.

The turn over is very sharp at FNAL, so, if the above is correct, F_{+-} is still very important at FNAL. It should then turn out that some spin dependent observables are <u>very</u> large at FNAL.

The whole picture will be tested by measuring the spin dependent parameters.

Which parameters will be big?

If ρ exchange is a sensible model then even though F_{+-} is large, it will have the <u>same phase</u> as F_{++}. Thus the polarizing power

$$P \propto \mathrm{Im}(F^*_{+-}\, F_{++}) = 0 \ .$$

But the spin rotation parameter

$$D_{ss} \propto \mathrm{Re}\,(F^*_{+-} F_{++})$$

will be large.

III. POLARIMETRY

Here, since the analyzing power of high energy elastic scattering is small, one would like to bring effective energy down. This can be done in inclusive scattering at fixed $p_\perp$ and x. For example, in the process a+d→b+X one may consider the contribution of the exchange process a+q→b+c and -q+d→X as shown in Fig. 7, the kinematics of which, for fixed x_b, x_c, $p_{\perp b}$ and $p_{\perp c}$ is:

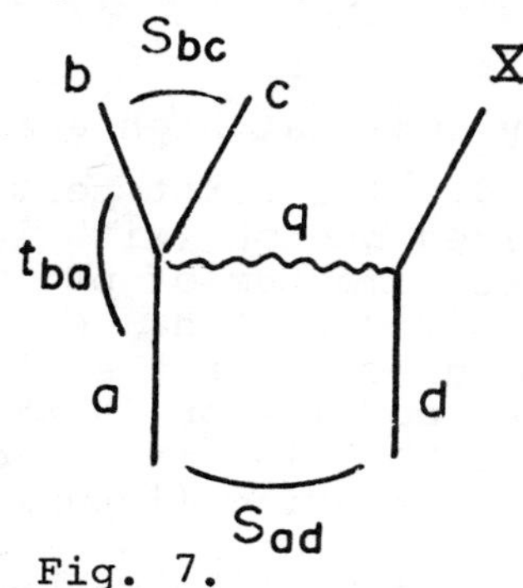

Fig. 7.

$$M_X^2=(1-x_b-x_c)\,S_{ad}$$

$$S_{bc}=(x_b+x_c)\left(\frac{m^2_{b\perp}}{x_b}+\frac{m^2_{c\perp}}{x_c}\right)-(\vec{p}_{\perp b}+\vec{p}_{\perp c})^2$$

$$t_{ba}=(x_b-1)\left(\frac{m^2_{b\perp}}{x_b}-m_a^2\right)-p^2_{\perp b}$$

and

$$q^2=(x_b+x_c-1)\left(\frac{m^2_{b\perp}}{x_b}+\frac{m^2_{c\perp}}{x_c}-m_a^2\right)-(\vec{p}_{\perp b}+\vec{p}_{\perp c})^2 .$$

Thus, at fixed x_b and x_c with $x_b+x_c<1$, the polarization will be energy independent as $S_{ad}\to\infty$ (if q+d→X has a constant cross-section and the polarization in the process q+a→b+c is non-zero). The process under consideration might be p+p→Λ+X, with the subprocesses p+π→Λ+K and π+p→X.

*Another process which may be considered is the use of the Coulomb production of $p\pi^\circ$ by incident protons on nuclei as a possible absolute polarimeter for the incident proton's polarization. Because of current conservation, the amplitude for the process shown in Fig. 8 is given by (except for factors of "i" and "-1")

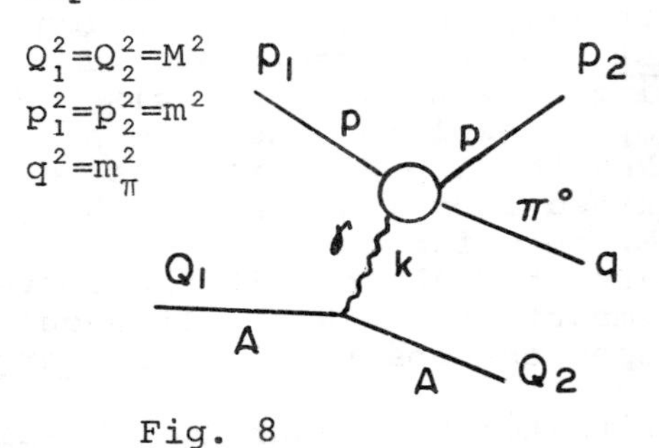

Fig. 8

$$\text{Amp} = \frac{Ze}{k^2} Q_1^\mu \mathcal{M}_\mu^{\gamma p\to\pi^\circ p}(p_1, k, p_2, q). \tag{1}$$

In the rest frame (projectile frame with the z axis along $-\vec{Q}_1$) of the incident proton, one can compare the photoproduction amplitude in (1) with on-shell photoproduction. Using current conservation we find in the projectile frame

$$Q_1^\mu \mathcal{M}_\mu = \frac{MP_L}{k_0^* m} \left\{\frac{E_L}{P_L} k^*_T \cdot \mathcal{M}^*_T + \left(k_0^* + \frac{E_L}{P_L} k_3^*\right)\mathcal{M}_3^*\right\} , \tag{2}$$

where P_L is the lab beam momentum ($E_L=\sqrt{p_L^2+m^2}$) and the '*''s are to remind you the expression is given in a specific frame.

The basic features of Coulomb scattering can be seen after expressing k_3^* in terms of k^2 and k_0^*. The result is

$$\text{Amp} = \frac{Ze}{k^2} \frac{M}{mk_0^*} E_L\left\{k^*_T \cdot \mathcal{M}^*_T + \left(\frac{mk^2}{2MP_L} - \frac{m^2 k_0^*}{P_L E_L}\right) \mathcal{M}_3^*\right\} \tag{3}$$

First, since $k^2_{min} \simeq -\left(\frac{S_{\pi p}-m^2}{2P_L}\right)^2$ is very small at high energies, the second term in (3) can be dropped. The first term describes both a transversely polarized photon (with linear polarization along k_T=lab. transverse momentum of $p\pi^\circ$ system) and a small longitudinally polarized photon. We ignore the latter as being small. We also ignore here the effects due to the nuclear form factors. The conclusion is that the dominant part of (3) in the Coulomb region is the on-shell photoproduction amplitude for scattering linearly polarized photons.

In practice this means when the orientation of $\underset{\sim}{k}_T$ is

*This discussion was given by Margolis and Thomas.

averaged over, the process in Fig. 1 can be used to analyze the transverse components of the incident protons' polarization. In general, by not averaging over $\underset{\sim}{k}_T$, the process will also analyze the longitudinal components of polarization.

OTHER CONTRIBUTIONS

*<u>Quark Model for polarizations in $pp \to \Delta^{++}n$ and $pn \to np$</u>. Remember that simple quark model predicts some relationship between the density matrix elements of the Δ in $pp \to \Delta^{++}n$ and those of K^{*0} in $K^+n \to K^{*0}p$ (Note the remarkable fact that this relates baryon-baryon scattering and meson-baryon scattering). These relations are extremely well satisfied by the data at 6 GeV/c.

Polarizations for $pp \to \Delta^{++}n$ and $pn \to np$ are large (40-50%) and show almost no energy dependence and for both reactions the cross sections have p_{lab}^{-2} behavior (except for the Δ reaction at ISR) which is usually interpreted as due to π exchange. The polarization data has no simple explanation in standard Regge models.

We propose a mechanism where the polarizations at large momentum transfer are due to strong spin dependent effects in quark-quark scattering. One has to look back to the old results in Møller (or Bhabha) scattering which exhibit this strong spin dependence (i.e. $\sigma_+/\sigma_- = 1/8$). This quark-quark component of the scattering gives the correct p_{lab}^{-2} behavior and is real. It should then interfere with a Reggeized π exchange which provides the necessary phase difference and would give a polarization independent of energy. Rough estimates of the spin dependent Møller cross sections show that we could easily get the right order of magnitude. More work remains to be done but this is an attractive idea.

**<u>Ask if available data on elastic $\pi^{\pm}p$ polarizations at p_{lab}=100 GeV/c has any piece coming from the tail of the Coulomb interference effect</u>. A way to detect that is to check isospin constraints and if they are not satisfied this would be an evidence for such an effect. Call σ_+, σ_- and σ_0 the differential cross sections for π^+p, π^-p and $\pi^-p \to \pi^0 n$ and P_+, P_- the two elastic polarizations. Isospin requires (after some reasonable approximations)

$$|P_+ - P_-| < \sqrt{2}\ \sqrt{4\rho - \Delta^2}$$

where $\rho = \dfrac{\sigma_0}{\sigma_+ + \sigma_-}$ and $\Delta = \dfrac{\sigma_+ - \sigma_-}{\sigma_+ + \sigma_-}$. It is clear on this inequality that if the tail is present it will enhance $P_+ - P_-$,

*This discussion was given by Margolis and Soffer.
**This discussion was given by Michel and Soffer.

but of course if the inequality is fulfilled one cannot rule out the effect.

The results for the most accurate points of $P_{\pm}$ are

(t)	$\|P_+ - P_-\|$	$\sqrt{2}\sqrt{4\rho-\Delta^2}$
0.19	6%	6.12%
0.25	5.3%	4.76%
0.35	3.4%	4.50%
0.45	0.4%	2.89%

which does not show a violation although we have not carefully calculated the errors.

We urge experimentalists to do this exercise whenever data makes it possible.

WORKSHOP ON POLARIZED PROTON BEAMS

ANN ARBOR, MICHIGAN, OCTOBER 18-27, 1977

REPORT OF WORKING GROUP ON ACCELERATOR PROBLEMS*

This workshop has concluded that there are many experimental elementary particle objectives which are uniquely achievable with polarized proton beams. These objectives require both higher beam energies and intensities than are presently available.

The polarized ion source working group has concluded that it looks quite practical to produce polarized H^- sources with output currents of several mA. The significance of this is best demonstrated by the fact that the ZGS operates at its space charge limit using a 6 mA H^- unpolarized ion source and charge exchange injection. Thus, polarized H^- ions offer the possibility of operating high energy synchrotrons at their "normal" intensity.

With interesting physics to be done and the high probability of adequate intensities for both fixed target and colliding beam machines, the question to be answered is whether or not it is possible to accelerate polarized protons and/or deuterons to energies above the presently available 12 GeV/c of the ZGS and to store beams in a colliding beam machine such as the ISR or ISABELLE. These questions were considered by the accelerator physics working group whose conclusions are discussed below.

FEASIBILITY OF MAINTAINING POLARIZATION

Polarization of a beam in a circular accelerator tends to be destroyed by the resonances described in the introductory talk at this workshop.[1] The following techniques are possible for coping with this problem:

---

*Y. Cho, W. Kubischta, B.W. Montague, A. Turrin; E.D. Courant† and L. Ratner, Co-Chairmen.

†Work performed under the auspices of the U.S. Department of Energy.

1. Referred to as I in this paper.

ISSN: 0094-243X/78/041/$1.50

A) Traverse resonances very rapidly. With pulsed quadrupoles, the tune ν can be changed rapidly and, therefore, the intrinsic resonances which occur when γ passes through

$$\gamma G \pm \nu = kP \tag{1}$$

[See I, eq. (8)] can be traversed fast enough to reduce the depolarization to a small value. This technique has been successfully employed at the ZGS. It also appears feasible for the Brookhaven AGS or CERN PS up to energies around 20 GeV.

However, this method does not apply to imperfection resonances. These are strongest for $\varkappa = \gamma G$ integral <u>near</u> the solutions of (1). At Argonne, these are overcome by carefully correcting magnet errors near each resonance. To do this, it is necessary to have a good measurement of polarization; thus, the results of the polarimeter workshops are very important here. It would seem possible, again, to make these corrections in the AGS or PS up to energies around 20 GeV.

B) If you can't lick 'em, join 'em. Enhance the resonances by going slowly, obtain spin reversal by the phenomenon of adiabatic passage. But, Argonne experience shows that this may be very difficult at the intrinsic resonances; probably because some particles have small amplitudes, and because synchrotron oscillations can cause particles to traverse the resonance repeatedly in one passage.

C) Use deuterons. The resonances for deuterons are 25 times as far apart in energy as for protons. Thus, the imperfection resonances are now 13 GeV apart rather than 0.52 GeV. Therefore, in the large accelerators (SPS, FNAL main ring, ISABELLE), it may be possible to accelerate polarized deuterons up to energies of around 300 GeV before encountering the most severe intrinsic resonance at $\gamma G = \varkappa \approx 22$ to 28.

THE SIBERIAN SNAKE - LONGITUDINAL POLARIZATION

Longitudinal Polarization - It may be desirable for some experiments to have the particles polarized longitudinally (helicity states) rather than transverse to the orbit plane. A technique invented by Derbenev, Kondratenko and Skrinsky[2] (Novosibirsk) makes this possible, and furthermore, avoids all depolarization resonances.

2. Ya S. Derbenev and A.M. Kondratenko, Novosibirsk preprint (1976), Derbenev, et al., submitted to Particle Accelerators (1977).

Consider a ring with two diametrically opposed long straight sections. Suppose in one of these, the beam is longitudinally polarized. In traversing the first semicircular arc, the spin precesses in the horizontal plane, and the angle between spin $\vec{S}$ and velocity $\vec{v}$ increases from zero to:

$$\alpha = \pi\gamma G$$

Now, if in the second straight section the spin can be made to precess through π around the beam axis, the angle between $\vec{S}$ and $\vec{v}$ then changes to $-\alpha$, and in the second semicircle it increases by α back to zero. Thus, the polarization is longitudinal on each traversal of the first straight section.

The precession by π in the second straight section can be accomplished, in principle, in several ways. The easiest to visualize is a solenoid (longitudinal field). In a solenoid, the precession rate is:

$$\frac{d\varphi}{d\ell} = (1+G)\,\frac{B}{B\rho}$$

therefore, the length of the solenoid would have to be such that $B\ell\ (1+G) = \pi\ (B\rho)$ or, for protons, $B\ell = 1.125\ B\rho = 35.2\ \beta\gamma$ kG-meters. This is impractical above, at most, a few GeV. However, a similar result can be achieved with transverse fields, taking advantage of the fact that the precession rate in a transverse field is γ times that in a longitudinal field and, therefore, independent of energy. A series of magnets with alternate transverse horizontal and vertical fields can achieve the desired net longitudinal precession. Each of the magnets precesses the spin through an angle of 90° with respect to the velocity and, therefore, must have a field length product of:

$$B\ell = \frac{\pi}{2}\,\frac{B\rho}{\gamma G} = 27.5 \text{ kG-m}$$

The following set of eight such magnets accomplishes the desired precession of π of the spin around the longitudinal axis, and at the same time restores the orbit to its original direction with zero net displacement.

Seven magnets precessing the spin successively through $\pi/2$ around the axes x, z, -x, -x, -z, x, -z, followed by a gap of the length of two magnets, and an eighth magnet precessing around +z. The first six magnets already accomplish our desired precession, but they also deflect the orbit by an angle $\theta = \pi/2(\gamma G)$ each.

In the vertical direction, the orbit is offset by $2\ell\theta$ by the first 3 magnets and restored by the second three; in the horizontal direction the set of six moves the orbit by a total offset of $3\ell\theta$. The last two magnets with the gap 2ℓ between them restore this offset to zero.

With 40 kilogauss, each magnet has to have a length of 0.69 meters, and the whole "wiggler" or "Siberian Snake" has to be 6.9 meters long. This is too long for the existing "small" accelerators such as the AGS or CERN PS, but would fit easily into the SPS, FNAL main ring, or ISABELLE.

The peak vertical excursion in the "snake" is 5.7 cm vertically and 8.5 cm horizontally at 20 GeV, decreasing with $1/\gamma$ for higher energies. Thus, the "snake" seems compatible with reasonable aperture requirement for proton energies above around 20 GeV. (It should be remarked that this scheme is out of the question for deuterons since their γG, at any energy, is 25 times less than for protons, so that a unit magnet would require 685 kG-m.)

An alternative "Siberian Snake" scheme replaces the first six magnets of the snake with three magnets, with $B\ell$ = 55 kG-meters each (precession angle π), fields oriented at angles -120°, 0, and $+120^o$ from the horizontal. In this case, the invariant spin in the opposite straight section is <u>radial</u>, rather than longitudinal.

Why is the "Siberian Snake" scheme immune to depolarizing resonances?

In any magnetic field, with or without magnet errors, there exists a closed orbit; along that orbit there exists a periodic spin vector $\vec{s}_o(\theta)$ which repeats on every revolution (its negative $-\vec{s}_o$ also does). The actual spin precesses around $\vec{s}_o(\theta)$.

Now for the "snake" configuration, the frequency of this precession around $\vec{s}_o(\theta)$ is just half the orbit frequency in the absence of field errors. Therefore, if field errors exist, any given field error has equal and opposite effects on the spin on successive revolutions, giving a cancellation. A more sophisticated calculation using the transformation of 2-component spinors by 2×2 SU2 transport matrices, shows that the <u>eigenvectors</u> of the matrix for a whole revolution are insensitive to small perturbations unless the <u>eigenvalues</u> of the SU2 matrix are degenerate, equal to +1 or -1 (recall that the precession angle of a <u>spinor</u> is just half that of the corresponding <u>spin</u>). For the "Siberian Snake" the SU2 matrix for the whole revolution is just $i\sigma_2$ or $i\sigma_1$ (depending on which variant is chosen), <u>eigenvalues</u> of which are $\pm$ i, as far from degeneracy as possible. This fact is independent of the precession angle in the arcs and, therefore, of energy; thus, resonances do not arise.

SCENARIOS FOR POLARIZED ACCELERATORS

The following scenarios may be suggested for the production of high energy polarized beams at Brookhaven, CERN or FNAL:

1. Accelerate vertically polarized protons in the source accelerator (AGS, PS, FNAL booster) to around 20 GeV (10 GeV at FNAL), killing the intrinsic resonances by ν-jumps and the imperfection resonances by careful compensation, as is now done at ANL. CPS or AGS 20 GeV beams can be used directly for fixed-target experiments; CPS beam could also be injected into the ISR. If improved sources, described in the source workshop, turn out to be possible, intensities and luminosity at the ISR may be 10% or more of those now obtained with unpolarized beams. One might also inject into ISABELLE from the AGS and let the polarized beams circulate at 20 GeV on 20 GeV, to avoid depolarizing resonances in ISABELLE.

2. Install a "Siberian Snake" in one of the long straight sections of the SPS, ISABELLE or the FNAL main ring or energy doubler. Transfer the beam from the preaccelerator into the large ring, rotating the spin to the longitudinal direction on the way. Accelerate to top energy in the large ring.

Alternatively, accelerate polarized deuterons in the pre-accelerator, transfer to the large ring and continue to accelerate. Resonances will arise, but can probably be compensated up to the major one at $\gamma G = \nu$, which occurs at about 290 GeV in ISABELLE or the Fermilab accelerator and at 370 GeV at the SPS.

It should be noted, however, that this implies accelerating the deuteron beam in ISABELLE through transition (occurring at 35 GeV). Preliminary estimates indicate that this should not be too difficult, in spite of the slow rate of acceleration, because the acceleration frequency for ISABELLE is very low and, therefore, the longitudinal space charge effect will be relatively weak.

OTHER TOPICS COVERED IN WORKSHOP

Depolarization during storage - It appears plausible that depolarization during storage is governed by the same dynamics as in the case of electron beams, if the damping lifetime of the electron beam is replaced by the growth time of betatron oscillations due to scattering and/or instabilities.[3] If the beam is reasonably far from a resonance, this leads to a polarization lifetime of the same order as the beam lifetime.

Rederivation of Spin Equation - A. Turrin presented a new derivation of the dynamics of depolarization, which shows that the Froissart-Stora result may be modified so as to give complete spin-flip if the time program of passage through resonance is suitably altered.

Possibilities of Further Studies - Argonne Laboratory, with its existing polarized proton accelerator, is in a unique position to undertake experimental studies of storage and acceleration of polarized beams. The working group recommends that Argonne attempt the following:

1. Further study of depolarization, if any, during storage. The ZGS may operate with a flattop of 5 minutes at 6 GeV/c; this should enable them to measure depolarization times with a precision of the order of 10^{-4} sec^{-1}.

2. Further study of adiabatic spin flip. Can the results be improved with a careful control of synchrotron oscillations?

3. A 50 MeV storage machine is being proposed for an investigation of polarized beams in a strong focusing accelerator. One could investigate: depolarization during storage and resonance crossing, internal polarimeters, stochastic cooling, and some aspects of the "Siberian Snake" dynamics.

---

3. B.W. Montague, Report DESY 77/31, May 1977.

POLARIZED ION SOURCES AND LOW ENERGY COLLECTOR RINGS

H. F. Glavish
ANAC Inc. P.O. Box 7453, Menlo Park, CA 94025

ABSTRACT

This paper reviews the basic design of present day polarized ion sources, including the new colliding beam method. A summary of the workshop discussions of E. F. Parker, E. P. Chamberlin, W. Kubischta, and the author, are included. The discussions centered about ways to improve the beam intensity. A possible 1 mA pulsed, polarized negative ion source has been formulated. The CERN polarized gas jet target has also been considered.

Low energy collector rings, studied in conjunction with the accelerator group, have been found to offer little advantage, especially in light of the advances now being made with source technology.

POLARIZED ION SOURCES

A beam of polarized ions can be produced by ionizing a beam of polarized neutral atoms. Two methods exist for making a polarized neutral beam. They are based on quite different physical principles:

1. The Atomic Beam Method: a ground state $1S_{\frac{1}{2}}$ atomic beam is formed by gaseous effusion of atoms through a nozzle into a vacuum. The beam is polarized by the Stern-Gerlach method. i.e. by passing it through an inhomogeneous magnetic field.
2. The Lamb-shift Method: in this method a fast metastable $2S_{\frac{1}{2}}$ atomic beam is formed by neutralizing a 500 eV ion beam in a gas vapor canal. The neutral beam is then polarized by making use of the Lamb-shift between the $2P_{\frac{1}{2}}$ states and the $2S_{\frac{1}{2}}$ states.

Both types of source work with hydrogen and deuterium, but extension to other atomic species is possible. Extensive review articles in the literature[1-4] deal with the operating principles of both types of source, and the advances and refinements that have been made up until 1975.

The techniques used to ionize the neutral beam include electron bombardment, charge transfer and charge exchange. It is important that the ionization process is selective in order to minimize the production of unpolarized background ions. In the case of the atomic beam method the background ions can arise from residual gas in the ionizer vacuum system, and also from the molecular beam component, which is always present along with the atomic beam. In the case of the Lamb-shift method, the unpolarized background can arise from ground state $1S_{\frac{1}{2}}$ atoms, which are always present along with the metastable atoms.

A schematic diagram of a typical atomic beam source, producing polarized positive ions by electron bombardment of the neutral atomic beam, is shown in Fig. 1. The Lamb-shift source, with an argon gas cell ionizer, is shown in Fig. 2. The actual mechanical configurations of typical atomic beam and Lamb-shift sources are shown in detail in Figs. 3 and 4.

ISSN: 0094-243X/78/047/$1.50

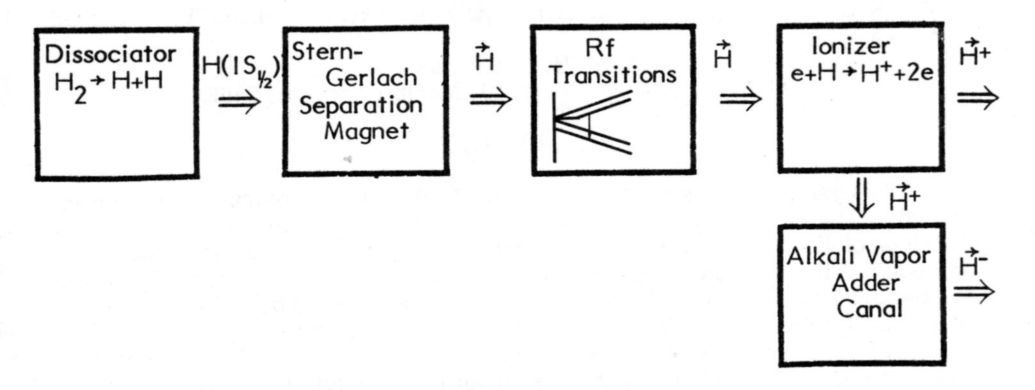

Fig. 1. Components of an atomic beam source. Atoms are generated by dissociating molecules in a low pressure dicharge. The atoms flow through a nozzle and form a supersonic beam. The beam is polarized by passing it through an inhomogeneous magnetic field, usually produced by either a sextupole or a quadrupole. Rf transitions are induced between the hyperfine states of the atoms to enable different nuclear polarization states to be selected, and to provide rapid spin reversal. The polarized atomic beam is then ionized. Positive ions are produced by electron bombardment. An adder canal can be used to convert these to negative ions. Alternative ionizing schemes can be used as in the colliding beam source (see Figs. 8, 9, and 10).

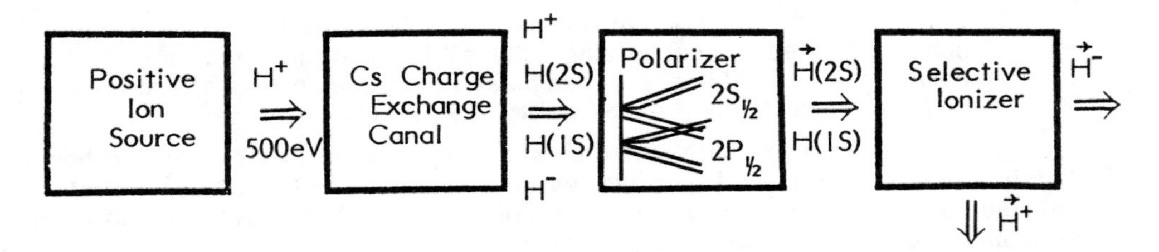

Fig. 2. Components of a Lamb-shift source. Protons, usually obtained from a duoplasmatron, pass through a cesium charge exchange canal. Approximately 10 to 20 % of them are converted into metastable $2S_{\frac{1}{2}}$ atoms. Charged components emerging from the cesium canal are deflected out of the neutral beam by electrostatic deflector plates. The metastable beam is polarized either by static fields, or static fields in conjunction with rf fields. The fields induce transitions from a $2S_{\frac{1}{2}}$ state to a $2P_{\frac{1}{2}}$ state, which then spontaneously decays to the $1S_{\frac{1}{2}}$ ground state. The metastable beam, which is now polarized, is ionized in preference to the ground state beam, by a selective ionizer. Argon gas produces negative ions, and iodine vapor produces positive ions.

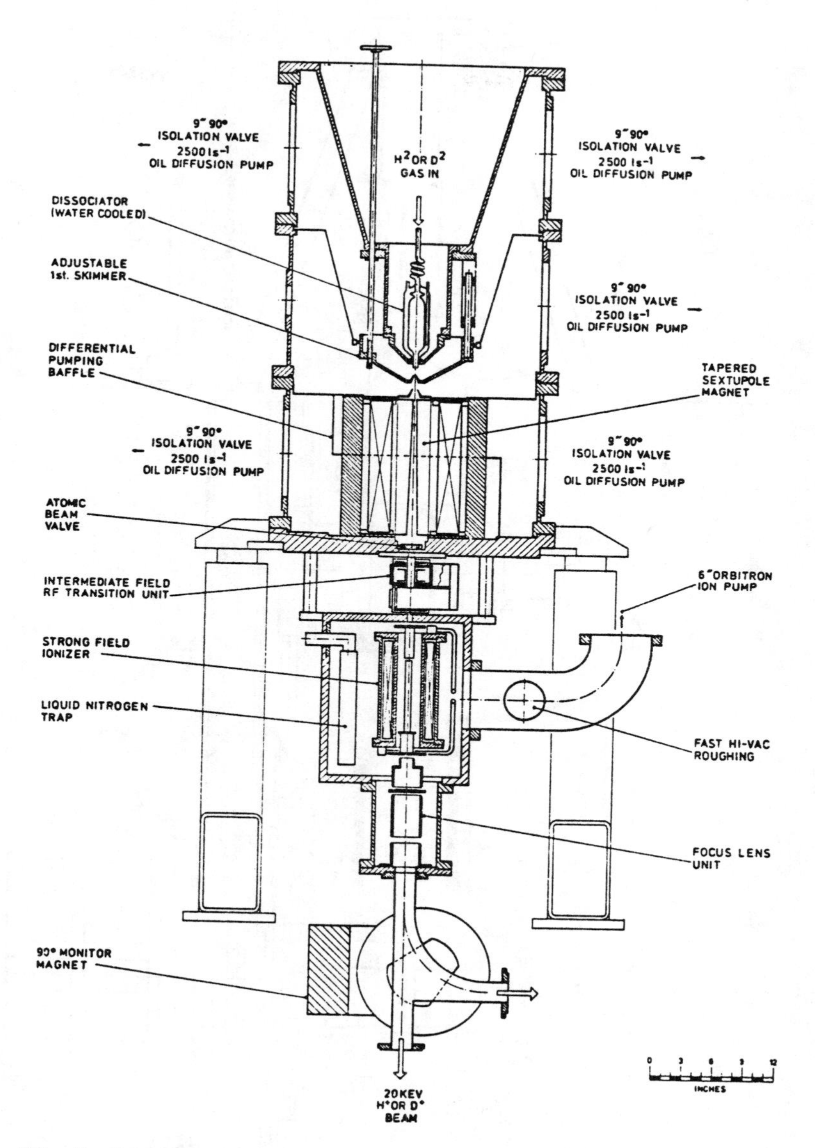

Fig. 3. Physical and mechanical details of an ANAC atomic beam source which produces polarized positive ions by electron bombardment.

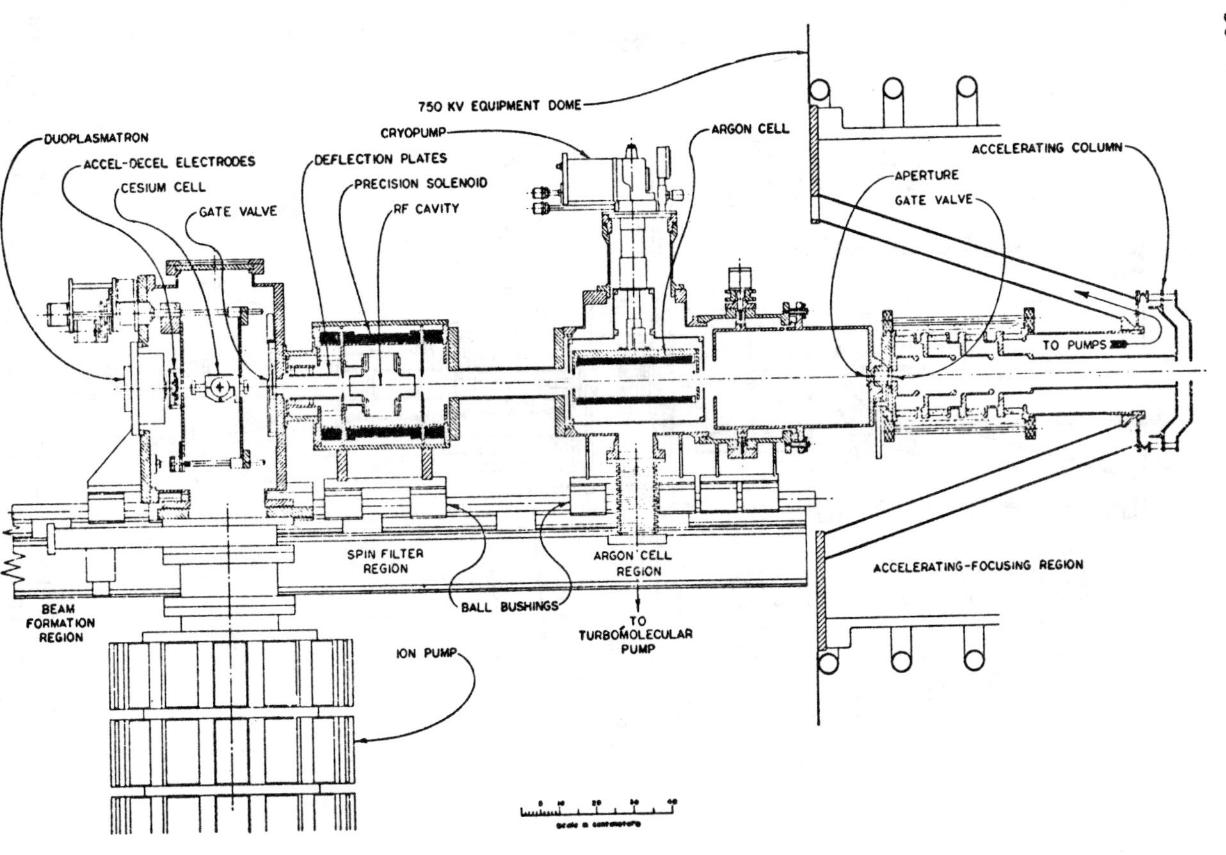

Fig. 4. Details of the Lamb-shift source used at LAMPF. The operating characteristics of this source are given in Table 1 of the text.

BEAM INTENSITIES

The atomic beam method is the oldest scheme for making polarized ion beams. Useable sources, producing approximately 0.1 μA of polarized H^+ and D^+ beams, came into existence around 1960. Ionization was by electron bombardment. By 1965, beam currents increased to 1.0 μA, with the introduction of a more efficient ionizer[5] that used a magnetic field of 1 to 2 kGauss to confine the electrons in a long cylindrical volume through which the neutral atomic beam passed. By 1972, a beam current of 10 μA was reached, and by 1976 an advance to 60 μA was achieved.

On this basis one might predict another order of magnitude improvement by 1980. But the mere passage of time does not lead to higher beam currents. In fact, all of the advances have resulted from quite intensive development programs, of refinements in design and testing, over a period of time.

Beam emittence can be as important as beam current. For electron bombardment ionizers, the beam emittence is determined almost entirely by the strength of the magnetic field in which the ions are formed[6]. For fields of 1 to 2 kGauss, the emittence is typically in the range of 35 to 60 $mm.mrad.MeV^{\frac{1}{2}}$.

The atomic beam method has also been used to make polarized negative ion beams, H^- and D^-, by passing the positive ion beam through an alkali vapor adder canal[7]. The conversion efficiency is approximately 5 %. The maximum negative beam current that has been obtained in this way is approximately 300 nA, derived from a 6 μA positive ion beam[8].

Ionization by electron bombardment is not the only way of producing polarized beams in an atomic beam source. It is also possible to use a colliding beam technique, involving charge transfer and charge exchange reactions, as proposed by Haeberli[9] in 1968. Quite recently such a source has been built and tested at the University of Wisconsin. The reaction

$$H^o + Cs^o = H^- + Cs^+$$

is used to produce polarized negative ions by colliding a 40 keV Cs^o beam with the polarized, thermal energy, atomic beam. A beam current as high as 3 μA has already been achieved[10], which at this time is a world record for polarized negative ions. The colliding beam technique promises significant advances in the near future, particularly for polarized negative ions. In fact, at this workshop, a scheme has been discussed in some detail, which in principle should produce a 1 mA polarized H^- or D^- beam. The colliding beam reaction that would be used is

$$H^o + D^- = H^- + D^o$$

with a high current (unpolarized) D^- beam generated from a surface plasma ion source of the type developed by Dudnikov[11] and built by others[12,13]. The source would operate only in pulsed mode, but for synchrotrons this is not a disadvantage. Furthermore, for synchrotrons, the hydrogen dissociator of the atomic beam source can be pulsed too, since as shown by Parker[14], this doubles the beam current.

Lamb-shift sources first became operational 5 to 10 years after the atomic beam sources. With an argon gas ionizer, the Lamb-shift source produces polarized negative ions directly, and with a small beam emittence.

For this reason such sources have been popular with electrostatic tandem accelerators, which have a very small acceptance. Probably as much development work has gone into Lamb-shift sources as into atomic beam sources. However the rewards have not been as spectacular. The first Lamb-shift sources produced a few hundred nanoamps of beam[15] but only a factor of 2 to 3 improvement has been achieved since then. The present limitation appears to be associated with space charge forces and quenching of the metastable atoms. These limitations have been discussed at some length at this workshop.

In summary, for synchrotrons, pulsed operation of an atomic beam source, producing polarized negative ions by a colliding beam technique, promises substantial increases in beam current in the near future. In this connection, it should be mentioned that negative ions can be injected more efficiently into synchrotrons than positive ions, since they can be stripped on injection. The source session at this workshop concluded that in terms of the number of particles that could be injected into a synchrotron ring, 1 μA of negative ions is worth 10 to 20 μA of positives, and perhaps even more, assuming that the space charge limit for the ring is not reached. The colliding beam source at the University of Wisconsin already produces 3 μA of negative ions, in dc mode. With a pulsed dissociator this projects to 6 μA, which is equivalent or better than what is currently achieved with the polarized positive atomic beam source presently operating on the ZGS. Apart from the relative merits of positive versus negative ions, it seems entirely likely that in the not too distant future, high energy physicists will have the opportunity of working with polarized beams of intensity more comparable to that of unpolarized beams.

SPECIFIC WORKSHOP DISCUSSIONS

Lamb-shift Source: E. P. Chamberlin reported on the LAMPF polarized ion source facility. The source is mounted in the dome of a 750 kV Cockcroft-Walton injector. It delivers H^- ions, which are accelerated in the 800 MeV linac simultaneously with a high intensity, unpolarized H^+ beam. Future operation may require a polarized H^+ beam, if a high intensity H^- unpolarized source is installed. The design objectives for the LAMPF source are set out in Table I. At the time of this workshop the source was operating reliably, at about 50% of the design beam current.

The mechanical layout of the source is shown in Fig. 4. For producing positive ions, iodine vapor would be substituted for argon in the ionization gas cell. Some of the design features of this source are aimed specifically at conducting parity violation experiments, which require rapid spin reversal without polarization distortion over the cross section of the beam, and without intensity modulation[16].

The intensity limitations of a Lamb-shift source were discussed in a general way. The positive ion beam from the duoplasmatron and associated focusing lens, quickly diverges unless neutralized by electrons or negative ions. In any case, in practice, it seems that the positive ion beam entering the cesium cell is limited to 1 to 5 mA. The other limitation, which is perhaps more fundemental in nature, is quenching of the metastable atoms by the charged component of the beam emerging from the cesium canal. The quenching action arises from the intrinsic electric field in the charged beam. The charged beam component is electrostatically deflected away, but this cannot be done too quickly for otherwise the field from the deflection plates will itself induce quenching[4].

Table I. The LAMPF polarized source design objectives.

Beams	H^+ or H^-
Beam emittence	0.02 cm-mrad normalized
Peak current	0.5 μA
Duty factor	6% (12% in future)
Average current	30 nA
Expected linac current	20 nA
Polarization	> 85%
Spin selection system	Nuclear spin filter
Spin reversal system	1) B field reversal in spin filter and argon cell for slow reversal 2) Resonant passage scheme for fast reversal
Spin precessor	Crossed E and B field analyzer on 750 kV beam transport line

Atomic Beam Source: As noted earlier, significant advances have been made with the intensity that can be achieved with a positive ion atomic beam source. Prior to 1975 the maximum beam current was 10 μA, achieved with a commercial source built by ANAC[17]. The first indication that the beam current might be enhanced substantially, came from the work at ANL, where it was demonstrated that the beam from an ANAC source could be enhanced a factor of 2 to 3 by pulsing the dissociator. Attention was then focused on the various components of the source to see if they were optimally designed. The conclusion was that they were not.

A program to build a better source was undertaken by CERN in conjunction with ANAC. Meanwhile, some new improvements were also incorporated in the ZGS source. By and large, both programs were successful. The CERN source produces a 60 μA beam of polarized H^+ or D^+ ions in dc mode, and over 100 μA when the dissociator is pulsed[18]. The source on the ZGS now delivers up to 70 μA pulsed, and the dc beam current, although not measured directly, is estimated to be approximately 30 μA[19].

The improvements are attributed to the following:

1. Shorter drift space from the nozzle to the sextupole entrance, improving the solid angle of acceptance by the sextupole at the nozzle.

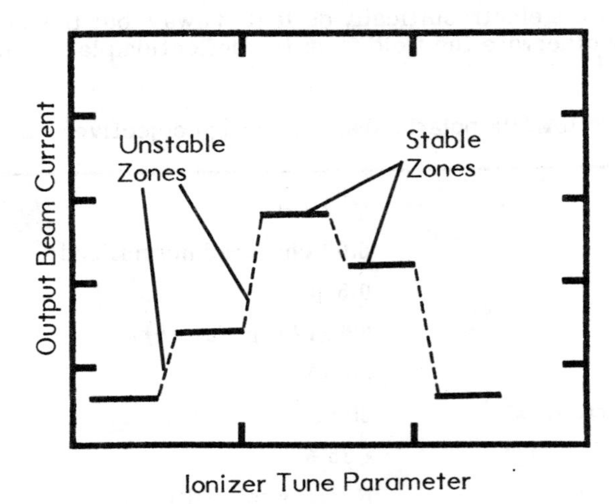

Fig. 5. Qualitative behaviour of the CERN strong field ionizer. The output current jumps from one stable zone to another as the ionizer is tuned.

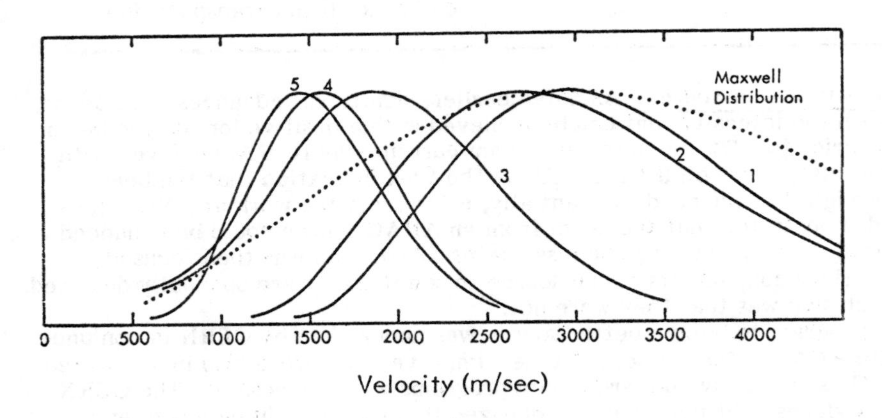

Fig. 6. Velocity distribution measurements made on the CERN atomic beam source. The experimental curves are: 1. H atoms with rf dissociator in pulsed operation; 2. H atoms with rf dissociator in continuous operation; 3. H_2 molecules with rf dissociator off; 4. H atoms with microwave dissociator in pulsed operation; 5. H atoms with microwave dissociator in continuous operation. All of the distributions are narrower than the Maxwell distribution (dotted curve), suggesting the beams are supersonic with a mach number of 1.5.

2. Improved vacuum conductance in the region of the nozzle and first part of the sextupole.

3. Addition of a second sextupole to optimize the focusing into the ionizer.

4. A longer ionizer, with magnetic field contouring to produce a higher, stable, electron space current.

An interesting phenomena was observed with the CERN ionizer. It was found to have a number of stable modes of operation, of essentially a discrete nature. It was not possible to pass from one mode to another in a continuous way. The transition occurred suddenly, as a jump. What might typically happen is shown qualitatively in Fig. 5. The tune parameter could be a magnetic field shape, or an electrode potential acting either on the electron beam or associated with the extraction region. As the tune parameter was varied, the output ion yield would resist change, until ultimately forced to jump to a new level. When the ion yield was examined with a scope, then at the point where a jump was about to be made from one level to another, it was observed that the ionizer was actually oscillating between the two adjacent stable levels, with a well defined square wave frequency, of the order of several kHz.

E. Parker reported on the experiments at ANL with an omega shaped filament of the type tried on one of the first strong field ionizers ever built[5]. Higher efficiencies were observed, but filament stability was a problem, even with a more heavily constructed filament.

W. Kubischta reported on the velocity distribution measurements made on the atomic and molecular beams of the CERN source. The measurements bring out three important points, of which little was known about until now.

1. Both the atomic and molecular beams are slightly supersonic, with a mach number of approximately 1.5.

2. When the dissociator is pulsed, the mean beam velocity increases fractionally.

3. A low temperature atomic beam can be formed, with a high dissociation degree, using a liquid nitrogen cooled microwave dissociator. (Prior to these measurements, there has always been a concern that even if a liquid nitrogen cooled dissociator is made to work, the atoms themselves may not be cooled to a low temperature, because of the limited heat transfer between the dissociator gas volume and the cooled surrounding jacket of the discharge tube.)

Some of the data associated with these measurements are shown in Fig. 6.

As mentioned earlier, since a synchrotron is a low duty cycle device, it is possible to operate the dissociator in a pulsed mode, and enjoy the benefits of a factor of 2 to 3 enhancement in beam current. This enhancement was first observed on the ZGS source, but has also been verified onthe CERN source. The enhancement observed on the CERN source was slightly less than a factor of 2, thought to be a result of two things: first the gas pulsing was probably not optimum as a result of a quite long response time of the gas feed valve to the dissociator. Second, some of the improvements incorporated in the CERN source, in particular the improved pumping in the region of the nozzle and the entrance to the sextupole, probably benefited the dc beam intensity as well as the pulsed beam intensity.

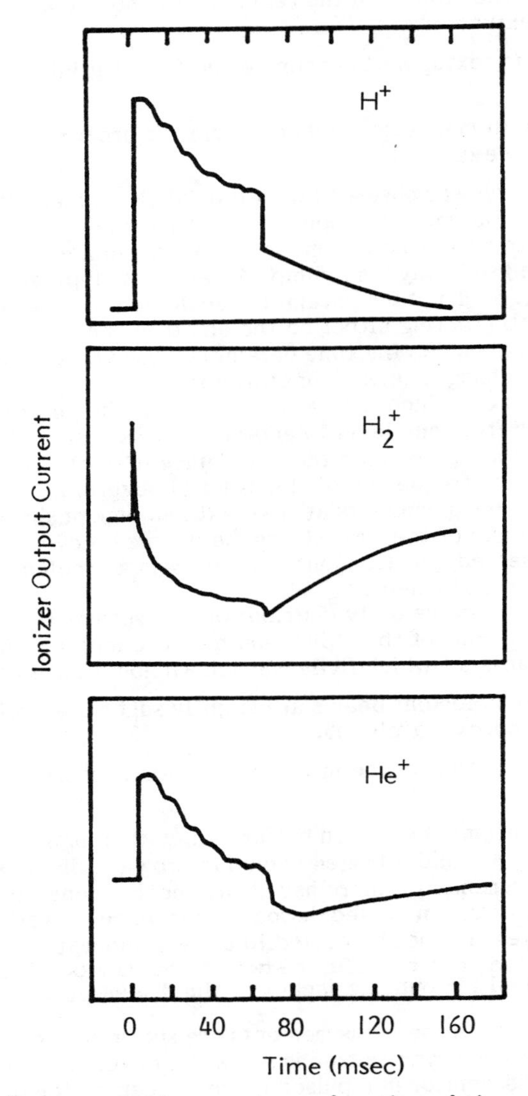

Fig. 7. Ionizer output current as a function of time when the rf dissociator is pulsed. The sextupole was off for all of the measurements. The dissociator is turned on at 0 msec and off at 70 msec. All currents have peak values twice their dc value.

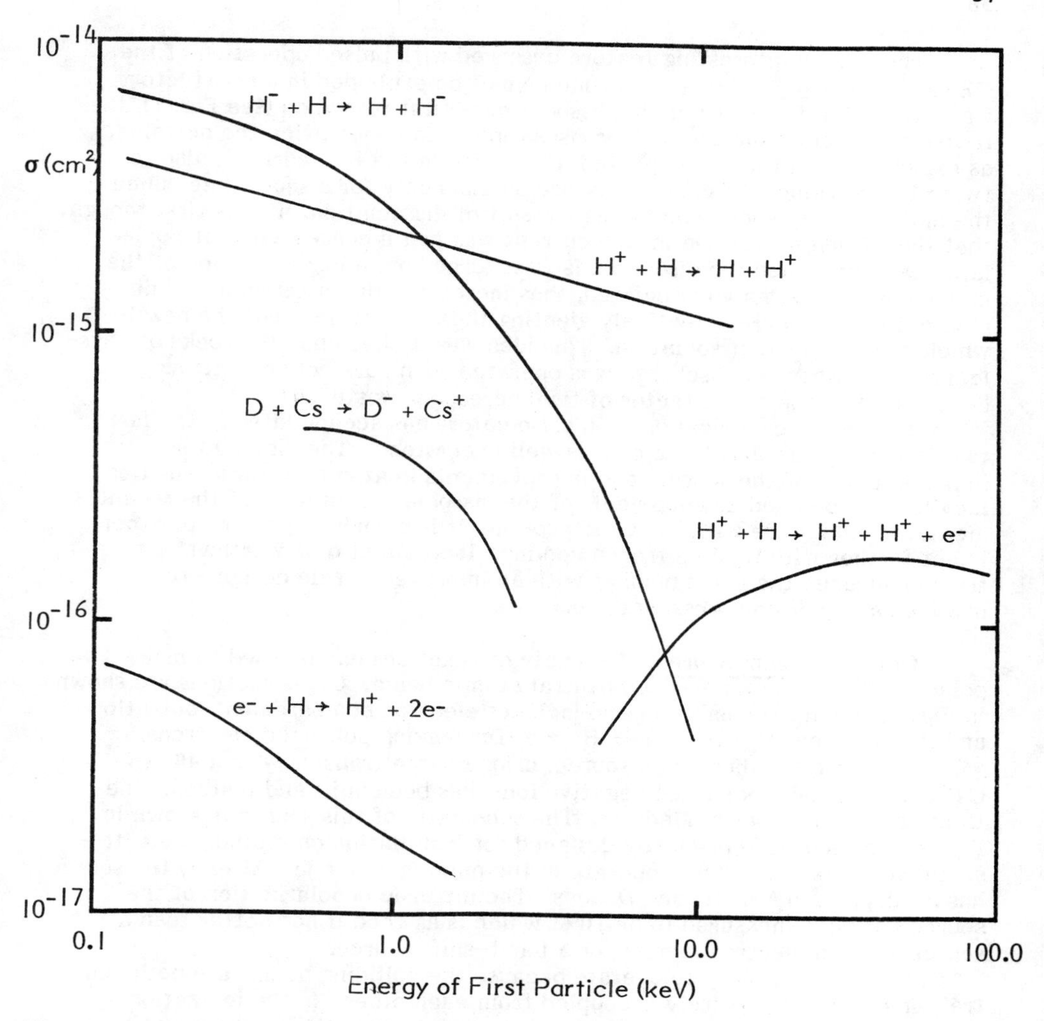

Fig. 8. Cross sections of various ionization processes. (Curves are approximate and should not be used as an absolute data source).

There is an interesting feature observed with pulsed operation of the CERN dissociator, which at this time cannot be explained in a satisfactory way. When the Rf power of the dissociator is first turned on (see Fig. 7), the H^+ beam current from the ionizer rises rapidly to about twice the dc value, as expected. What is less expected, is the rise in the H_2^+ current, also to twice the dc value. Of course this rise persists only for a short time, since the molecules are soon depleted as a result of dissociation. It was first thought that the sudden initial rise in H_2^+ current was just a consequence of Boyles law - namely - when the discharge is first turned on, a large fraction of the molecules are suddenly dissociated, thus increasing the pressure in the discharge tube and more effectively ejecting those molecules near the nozzle which have escaped dissociation. This idea was dashed upon the rocks of physical reality, when the discharge was operated with pure helium. Altough helium is monotonic, the factor of two increase persisted.

A lot more empirical data on dissociators has accumulated in the last two to three years, but little of it is well understood. The dissociator is an important part of the source, and improvements in atomic beam intensities inevitably depend on developments of the dissociator. In light of the recent measurements made with the CERN source, it is clearly important to experiment further with liquid nitrogen cooling. It would also be worthwhile to try and enhance the mach number with an improved nozzle design - for instance a higher compression ratio.

Colliding Beam Source: A variety of reactions can be used to make polarized ions from a polarized neutral atomic beam. Cross sections are shown in Fig. 8, which for completeness, includes electron bombardment ionization and the reaction $H^o + H^+ = H^+ + H^+ + e^-$ for making polarized electrons.

The first colliding beam source, using charge transfer with a 40 keV Cs^o beam to make polarized negative ions, has been built and tested at the University of Wisconsin, Madison. The schematic of this source is shown in Fig. 9. The source is primarily designed for installation on a tandem electrostatic accelerator, and will operate in the main in dc mode. Already the source has produced 3 μA of H^- and D^- ions. The intrinsic depolarization of the source has been measured to be 10%, which is as good if not better than a source with an electron ionizer, or a Lamb-shift source.

The source is easy to operate because the colliding beams are both neutral, and therefore entirely decoupled from each other. In the ionization region, electric and magnetic fields can be set up to optimize the negative ion extraction and polarization, without affecting either of the colliding beams. For tandem applications, this feature is important in regard to beam emittence. It has been found that the source operates equally well, with the same output beam current, regardless of whether the solenoid magnetic field in the interaction region is weak (a few Gauss) or strong (a few kGauss). But the beam emittence increases with magnetic field strength as noted earlier[6]. Now consider the case of deuterons. If the Rf transitions are induced in the atomic beam after the last sextupole, and just prior to the interaction region, a magnetic field of only 300 to 400 Gauss is required to realize all but a few percent of the available nuclear polarization. At this field strength, the beam emittence is less than 20 mm.mrad.$MeV^{\frac{1}{2}}$. For the case of protons, a higher magnetic field is needed of the order of 1500 Gauss, and consequently the

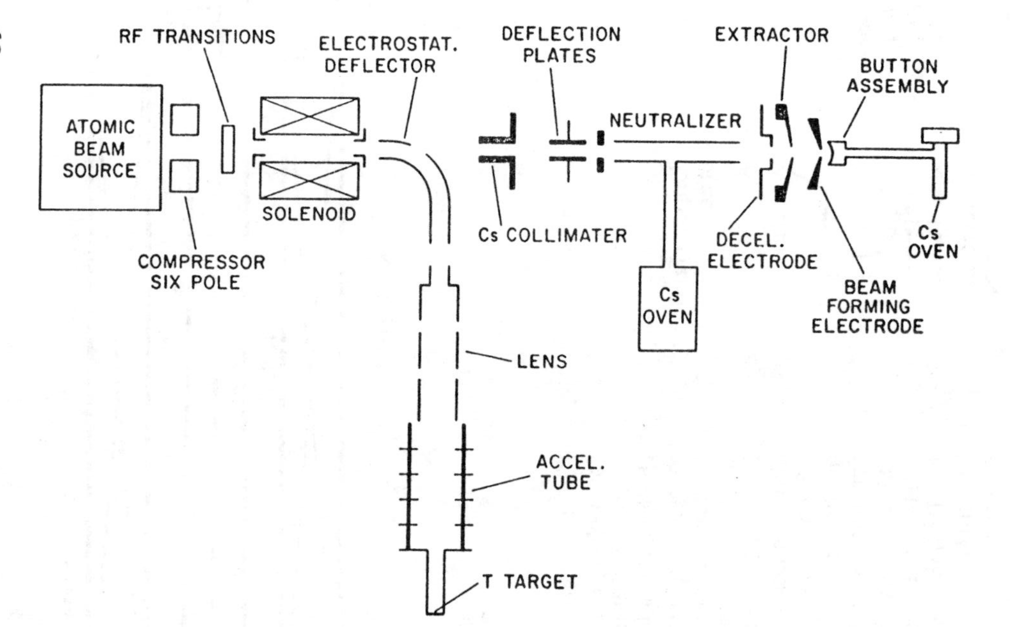

Fig. 9. The components of the University of Wisconsin colliding beam source. The source produces polarized H^- and D^- ions by colliding a 40 keV Cs^0 beam with the polarized neutral atomic beam. A Hot porous tungsten button produces Cs^+ ions by surface ionization of Cs metal. The Cs^+ ions are accelerated to 40 keV and then neutralized in a Cs vapor canal. The neutralized beam passes into a solenoid. Interaction with the atomic beam produces negative ions which are extracted and electrostatically deflected through 90°.

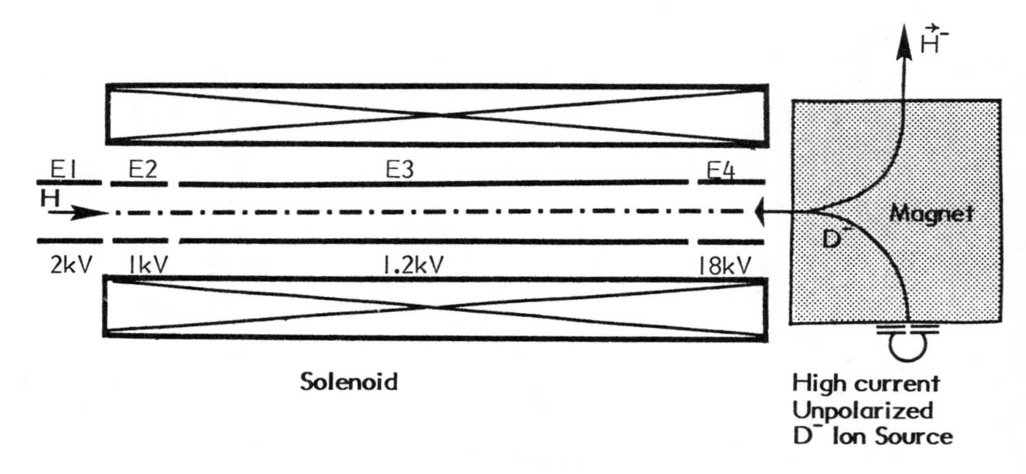

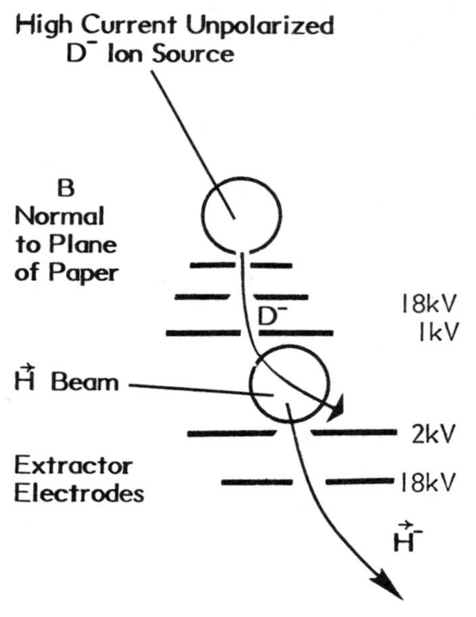

Fig. 10. Above: A possible arrangement for making a 1mA pulsed, polarized H^- beam. A 100 mA D^- beam collides with the atomic beam in a long cylindrical interaction region defined by E3. The D^- beam is decelerated on passing through the lens defined by E3 and E4. The potentials of E1 and E4 are higher than E3, and trap positive ions in the region of E3 which neutralize the space charge of the slow D^- beam. Polarized H^- ions formed in the interaction region are extracted by E4 and reflected by E2. Left: An alternative scheme where the D^- beam crosses the atomic beam. The interaction volume is smaller, but space charge neutralization may pose less of a problem. If a 1 A D^- source is used, this system would also have the potential to produce a 1 mA polarized H^- beam. Also the hardware would be more straightforward to set up, for preliminary experiments.

beam emittence is higher - perhaps 50 mm.mrad.MeV$^{\frac{1}{2}}$. However, since the colliding beam source can operate at very weak fields, just as well as at high fields, there is an alternative for protons. The Rf transition unit can be placed between two sextupoles. Full nuclear polarization is then achieved by ionizing in a field of a few Gauss for which the beam emittence has been measured as less than 10 mm.mrad.MeV$^{\frac{1}{2}}$.

Rapid spin reversal, with weak field ionization, can be arranged by placing a 2 - 4 transition unit between two sextupoles, and a weak field Rf unit after the last sextupole. Switching the weak field unit on and off reverses the spin. A scheme such as this would be of value in a parity violation experiment, since the beam intensity would not shift by more than 10^{-6} when the spin is reversed. The shift is an order of magnitude higher if ionization occurs in a strong magnetic field[20].

For a synchrotron, the Cs^o colliding beam source can operate with a pulsed dissociator, in which case it would deliver a beam of 6 μA. In a development sense, the Cs^o colliding beam source is in its infancy. Undoubtedly it will be improved in the near future. In particular, the Cs beam gun can be more optimally designed. Perhaps this would lead to a polarized, pulsed negative beam of 20 μA. In as much as polarized negative ions are worth 10 to 20 times more than polarized positive ions, at least for synchrotron injection, the Cs^o colliding beam source appears to offer more potential than an atomic beam positive ion source.

It is not necessary to restrict the colliding beam technique to just a Cs beam. It is also possible to use a colliding H^- or D^- beam to make polarized negative ions (see Fig. 8). The yield should be very high because the collision cross section is high, and also very high current pulsed hydrogen negative ion sources have recently been developed[11,12,13]. The current from these new surface plasma ion sources can be as high as 1 A of H^-, and only slightly less for D^-. The output beam from the sources typically have an energy of 15 to 20 keV. In order to realize a high cross section in the process:

$$H^o + D^- = H^- + D^o$$

the colliding negative beam must be decelerated to approximately 1 keV. The space charge associated with a high current beam can easily be neutralized by trapped ions of opposite charge if the beam is transported in a magnetic field. However, the moment the beam is decelerated, the decelerating electric field can sweep away the trapped ions. Thus special care must be taken. Two possible schemes, discussed at this workshop, are shown in Fig. 10. If this type of colliding beam source can be made to work, pulsed polarized currents as high as 1 mA are possible.

Polarized Gas Jet Target: A polarized neutral atomic beam may be used as a polarized target. When installed in an actual accelerator ring, data can be collected during most of the acceleration cycle. Thus, although the target has a low density, count rates are quite acceptable. In fact, when the advantages of a polarized gas jet target are considered, it has a figure of merit for many experiments which is better than a polarized solid target[21]. The advantages are:

1. High polarization - 80%.

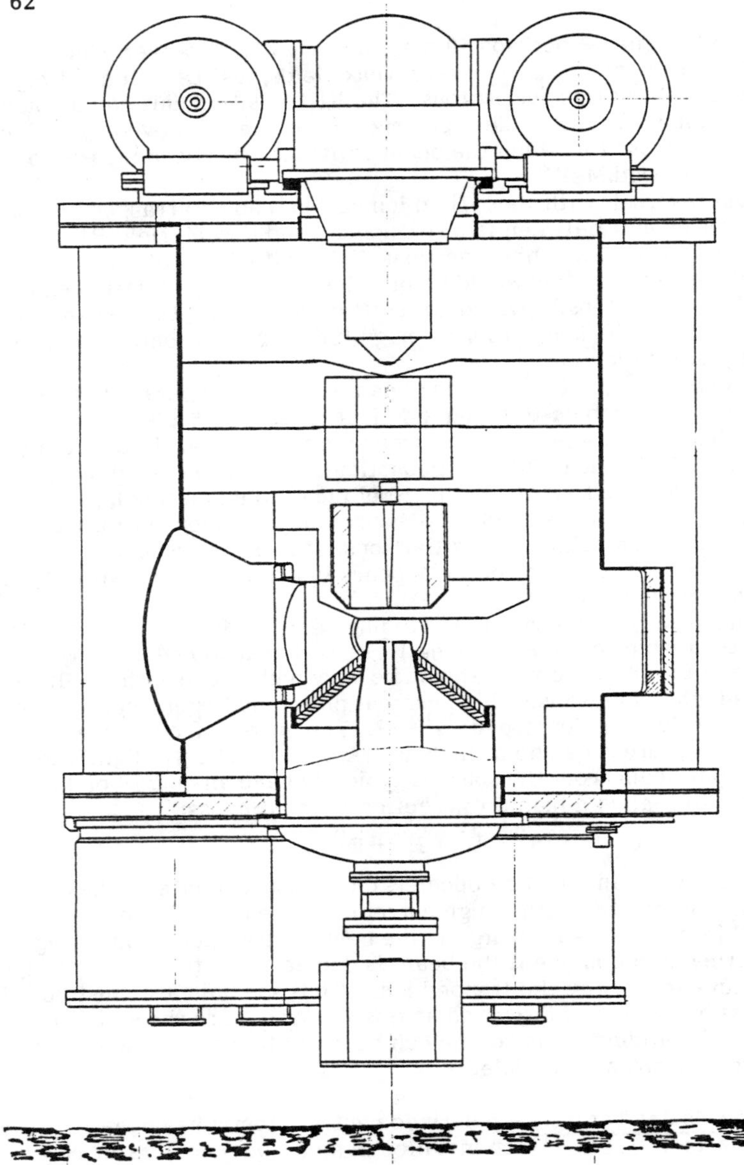

Fig. 11. The CERN design for a polarized gas jet target. Two sextupoles are used with an intermediate rf transition. Nuclear polarization is then achieved at the interaction region with just a weak magnetic field. Turbomolecular pumps are used throughout. The spent atomic beam is dumped in a cryopump. The system has an integral, ultra-high vacuum scattering chamber.

2. Rapid spin reversal, with any spin direction possible.
3. Point-like target, with negligibly small background.
4. Not damaged by high beam currents.
5. Inexpensive, low maintenance, and will operate continuously.

A polarized gas jet target has been designed at CERN. The features of this target were described at this workshop by W. Kubischta and L. Dick. The neutral beam system has two sextupoles with a 2 - 4 transition between them. Full nuclear polarization is then obtained, after the second sextupole, with just a weak magnetic field (see the discussion in the previous section). A weak field is an advantage in two respects: first, the coils producing the field may consist of just one or two turns, leaving the target open to be observed by detectors over a wide range of angles. Second, the weak field will not significantly perturb the beam of the accelerator.

The first sextupole has a diverging taper, as is usual for atomic beam sources. The second sextupole, however, has a converging taper to maximize the atomic beam density at the almost pointlike region where the accelerator beam crosses the atomic beam. The entire source has turbo-molecular pumps in order to meet the vacuum engineering requirements of modern synchrotrons, such as the SPS. The spent atomic beam is dumped in a cryopump which minimizes the background gas in the target region. Diagrams of the target are shown are shown in Fig. 11.

LOW ENERGY COLLECTOR RINGS

A synchrotron is a low duty cycle machine. As such, the injection time is only of the order of 100 μsec every synchrotron super period of 4 to 8 sec. Simple-minded thinking suggests that during the acceleration cycle, the polarized source could continue to operate, delivering its beam to a storage device. In this way there would be more particles available to transfer to the synchrotron over the injection period.

To implement such a scheme, there are practical points to consider, and physical laws to satisfy:

1. In a storage device for charged particles there is a limit to the number of particles that can be stored in a stable ensemble. This limit is set by space charge effects, scattering effects from a finite vacuum, the aperture or geometrical acceptance of the device, and the method or lattice which is used to stabilize the system against small perturbations.
2. It must be possible to successfully load into, and extract from the storage device. This involves the practical aspects of beam gymnastics and hardware limitations.
3. There must be emittence matching between the storage device and the accelerator ring. That is to say, whatever is stored and then extracted must be capable of being delivered to, and accepted by the accelerator ring.

The above mentioned points can be stated in various ways. But one thing is certain: while the concept of storage is simple to visualize, its implementation in practice is far from simple.

There is also something of a conceptual fallacy in the concept of low energy storage. Unless a synchrotron is restricted by equipment limitations at injection, the injection time is adjusted to fill the available synchrotron acceptance. This, of course, is all that can be done with a collector device; i.e., stack the beam in the storage device until the synchrotron acceptance is reached. Storage beyond this point would be useless. If there are equipment limitations, such as the linac pulse length being too short, storage rings could be used to circumvent these limitations. The practicality of this approach would depend upon the details of a particular situation.

A brute force approach is to install a large number of collector rings in the injection system. Provided that the acceptance of the synchrotron is not exceeded, the number of particles that could be injected into the synchrotron would be proportional to the number of collector rings. The improvement factor would be directly proportional to the cost, and the cost would be quite high. This scheme is neither clever nor elegant, and might be considered only if all else fails.

A situation where a collector would be of value will exist at the ZGS when a polarized H^- ion source becomes available. Using charge exchange injection (see next section), it would be possible to fill the ZGS to its space charge limit if the linac pulse width were long enough, the ZGS vacuum good enough, and the rf system could run at low frequencies for extended periods. Since these situations do not exist, the maximum charge that can be injected into the ZGS will be

$$Q = I \times 600\ \mu\text{sec}$$

where I is the linac output current and 600 μsec is the maximum pulse width capability of the ZGS linac.

A rapid cycling 500 MeV synchrotron is now operational at ANL. This machine, originally built to be a booster injector for the ZGS, uses the ZGS injector linac and is physically located where it could easily be coupled to the ZGS for polarized H^- injection.

This booster has an injection time of about 400 μsec using charge exchange injection, and eight pulses from this machine can be loaded into the ZGS at 500 MeV. Thus, using this machine as a collector ring would give the ZGS an effective injection time of 3200 μsec; more than a factor of 5 increase. Since this booster injects into the ZGS at 500 MeV, the vacuum and rf limitations of the ZGS are circumvented. The cost of installing a

500 MeV beam transfer line and injection system would be much less than improving the linac pulse length, the ZGS vacuum, and the ZGS rf.

NEGATIVE ION INJECTION

One way to make a beam brighter is to improve the source. Another way is to start off with negative ions, and then strip them to positive ions at injection into the accelerator ring. This is not a violation of Liouville's theorem if one considers the entire ensemble of stripper foil particles plus beam particles, and takes into account the composite nature of the negative ion beam particles. Successful experiments on stripping injection have been conducted at ANL, so the idea is a practical one. For polarized ions, the stripping process does not introduce depolarization, in the case of foil strippers, because the process occurs quickly in relation to the Larmor precession rate of an atomic spin. For a gas stripper, a longitudinal magnetic field of 1 or 2 kGauss will prevent depolarization.

With stripping injection the consecutive orbits do not need to be shifted to bring about optimum stacking with multi-turn injection. Thus, if 20 turn injection is possible with positive ions, at least 200 turn injection is possible with negative ions which are stripped at injection. Beam scattering from multiple passes through the foil is negligible at 50 MeV.

As a specific example, suppose a 500 μA source exists. This is not out of the question, using the colliding beam technique discussed earlier. If such a source is used in conjunction with the ZGS, an 600 μsec injection time would give 10^{12} polarized protons which is close to the unpolarized beam intensity.

REFERENCES

1. J.M. Dickson, Progr. Nucl. Tech. Instr. 1, 105 (1965).

2. W. Haeberli, Ann. Rev. Nuc. Sci. 17, 373 (1967).

3. H.F. Glavish, Proc. 3rd Int. Symp. on Polarization Phenomena in Nuclear Reactions (Univ. of Wisconsin Press, Madison, WI., 1971. Eds. H.H. Barschall and W. Haeberli), p. 267.

4. T. B. Clegg, Proc. 4th Int. Symp. on Polarization Phenomena in Nuclear Reactions (Berkhauser Verlag, Basel and Stuttgart, 1976. Eds. W. Gruebler and V. Konig), p. 111.
5. H. F. Glavish, Nucl. Instr. and Methods 65, 1 (1968).
6. G. G. Ohlsen, J. L. McKibben, R. R. Stevens, Jr., and G. P. Lawrence, Nucl. Instr. and Methods 73, 45 (1969).
7. W. Gruebler, P. A. Schmelzbach, V. Konig, and P. Marmier, Helv. Phys. Acta 43, 254 (1970).
8. B. A. MacKinnon et. al. Proc. Symp. on Ion Sources and Formation of Ion Beams (Brookhaven National Laboratory Report BNL 50310, 1971) p. 245.
9. W. Haeberli, Nucl. Instr. and Methods 62, 355 (1968).
10. H. F. Glavish, private communication.
11. V. G. Dudnikov, Surface - Plasma Source of Penning Geometry (IV USSR National Conference on Particle Accelerators, 1974).
12. P. W. Allison, IEEE Trans. Nuc. Sci. NS-24, 1594 (1977).
13. K. Prelec, Th. Sluyters, and M. Grossman, IEEE Trans. Nuc. Sci. NS-24, 1521 (1977).
14. E. F. Parker, N. Q. Sesol, and R. E. Timm, IEEE Trans. Nuc. Sci. NS-22, 1718 (1975).
15. B. L. Donnally, same as ref. 3, p. 295.
16. E. P. Chamberlin, R. R. Stevens, J. L. McKibben, IEEE Trans. Nuc. Sci. NS-24, 1524 (1977).
17. ANAC Ltd., P. O. Box 16066, Auckland, New Zealand.
18. W. Kubischta, B. A. MacKinnon, and H. F. Glavish, to be published.
19. E. F. Parker, private communication.
20. W. Haeberli, private communication.
21. J. Antille et. al., Report CERN/SPSC/77-71; SPSC/P 88, 26 August 1977, Geneva.
22. E. Bovet, R. Gouiran, I. Gumowski, K. H. Reich, Report CERN/MPS - SI/Int. DL/70/4, 23 April 1970, Geneva.

POLARIMETERS: A SUMMARY

J. B. Roberts, Jr.
Physics Department and T. W. Bonner Nuclear Laboratories
Rice University, Houston, Texas

ABSTRACT

The measurement of the polarization of the accelerated proton beam at the Argonne ZGS has become a straightforward procedure. However, measurement of the polarization of a proton beam accelerated in a very high energy accelerator or stored in a very high energy storage ring seemed to be a more formidable problem. Various possible methods of measuring the polarization of multihundred GeV proton and deuteron beams were studied. For proton beams several methods seemed to be feasible; in particular, those using known or calculable electromagnetic effects appeared quite promising. The choice of polarimeters for deuteron beams was more limited. Stripping the deuterons to make beams of polarized protons or neutrons whose polarization could then be measured by known techniques is a possibility. The measurement of inelastic spin effects at FERMILAB and SPS using polarized proton targets or polarized hydrogen gas jets to search for a good spin analyzer was encouraged.

INTRODUCTION

The advent of the polarized proton beam at the Argonne ZGS has permitted considerable improvement in technique for experiments which study the spin dependence of the strong interaction. Many new and exciting physics results have been discovered, and it seems compelling to extend the s range of these measurements up to ISR, FERMILAB, SPS, and ISABELLE energies. In order to maintain the polarization of the beam during acceleration, storage, and beam transport, and to perform experiments with the beam, measurement of the polarization of the beam with fairly high precision is necessary. Beam polarization measurements are made at the ZGS using the analyzing power for p-p elastic

ISSN: 0094-243X/78/067/$1.50

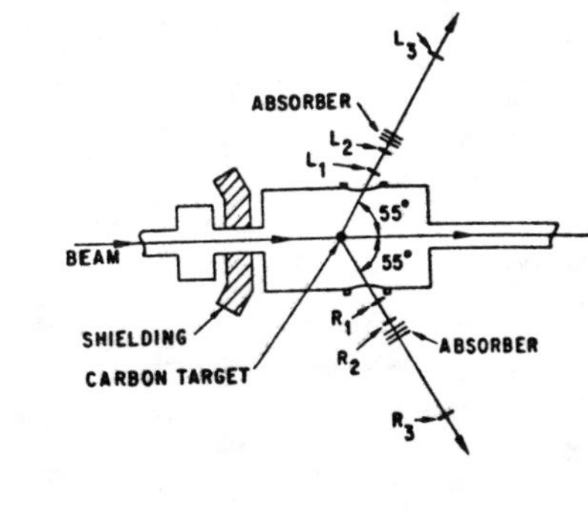

Fig.1a. 50-MeV Polarimeter showing the two symmetric 3-counter scintillator telescopes

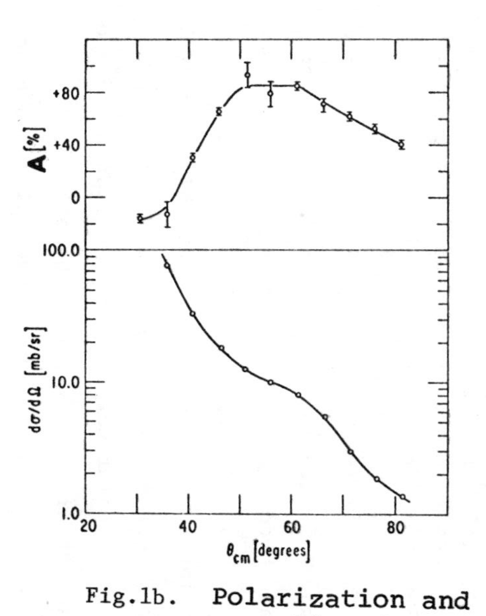

Fig.1b. **Polarization and cross section data for 50 MeV p-C scattering**

scattering (A_{pp}) which is measured using a polarized proton target whose polarization is measured by well-known NMR techniques. However, A has been found to be ∿1-2% for $|t|\lesssim 1$ in the momentum range 100 to 300 GeV/c.[1] Thus p-p elastic scattering in the diffraction peak is not a feasible process to use for beam polarization measurement. The size of A in elastic scattering at large t is not known so we must consider other reactions. Inclusive inelastic hadronic reactions, elastic scattering in the Coulomb region, and the diffractive production of Nπ systems were studied.

CURRENT POLARIMETERS

A schematic of the ZGS Linac polarimeter is shown in Fig.1a. This polarimeter uses p-Carbon elastic scattering at $\Theta_{LAB}=55°$ at 50 MeV, where the analyzing power is about 85% (Fig.1b). Range and time of flight are used to distinguish elastic events from the 4.43 MeV excited state of C^{12} and general lower energy background. The two identical arms, L and R, allow averaging over spurious asymmetries due to beam steering and position drifts between spin up and down pulses. A 2% relative measurement of the beam polarization can be made in less than a minute; there is an absolute error of 5% due to uncertainty in A_{pC}. Similar polarimeters might be used after the injector or pre-injector of any accelerator.

The beam polarization after extraction from the ZGS is measured with the elastic scattering polarimeter schematically shown in Fig.2.[2] The polarized beam tranverses the liquid H_2 target, and its polarization is measured by comparing the number of elastic events counted in the L and R spectrometers. The momenta and scattering angles of both forward and recoil protons are measured in each spectrometer, giving a very clean 4-constraint elastic trigger. The detectors are scintillation counters, so the asymmetry is read out instantaneously. The p-p analyzing power (A_{pp}) is measured very accurately by scattering the beam from the PPT, whose polarization is measured by NMR techniques. Some of the 6 and 12 GeV/c data for A are shown in Fig.3; the percentage errors are ∿3-5% for many of the data points, which in turn allows a beam polarization measurement of comparable accuracy.

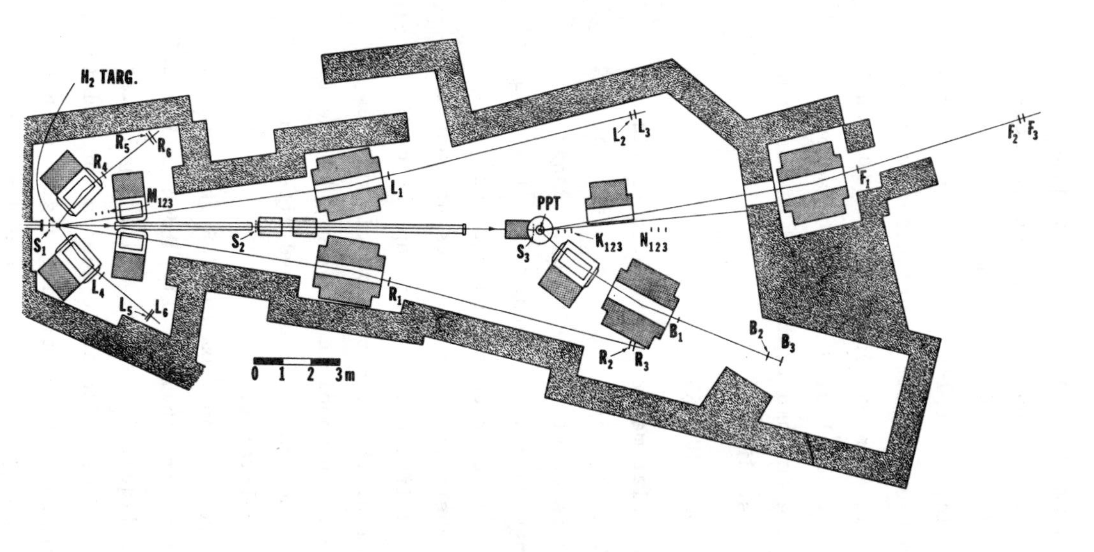

Fig.2. Layout of the high energy polarimeter and downstream polarized target. The polarized beam passes through the liquid H_2 target, and its polarization is measured by comparing the number of elastic events seen in the L and R spectrometers of the polarimeter. The beam then scatters in the polarized proton target (PPT) and the elastic events are counted by the F and B counters.

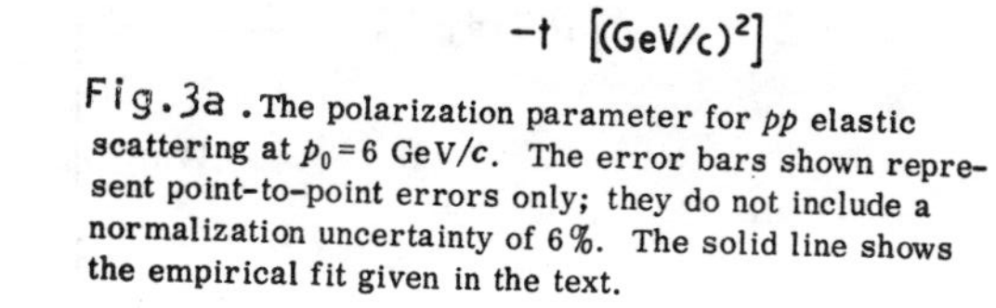

Fig.3a. The polarization parameter for pp elastic scattering at $p_0 = 6$ GeV/c. The error bars shown represent point-to-point errors only; they do not include a normalization uncertainty of 6%. The solid line shows the empirical fit given in the text.

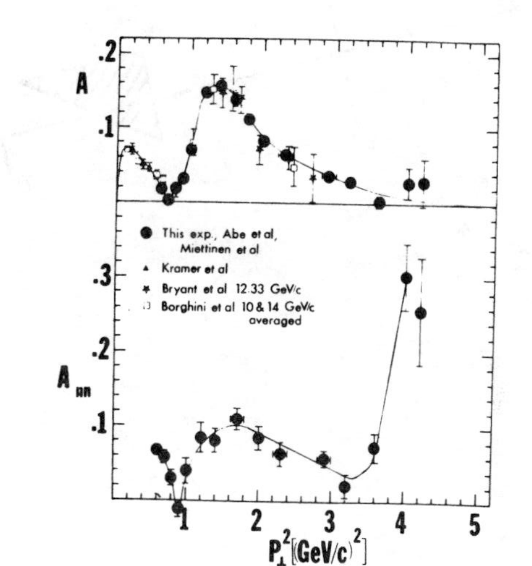

Fig.3b: The Wolfenstein parameters A and A_{nn} for p-p elastic scattering at 11.75 GeV/c are plotted against $P_\perp^2$.

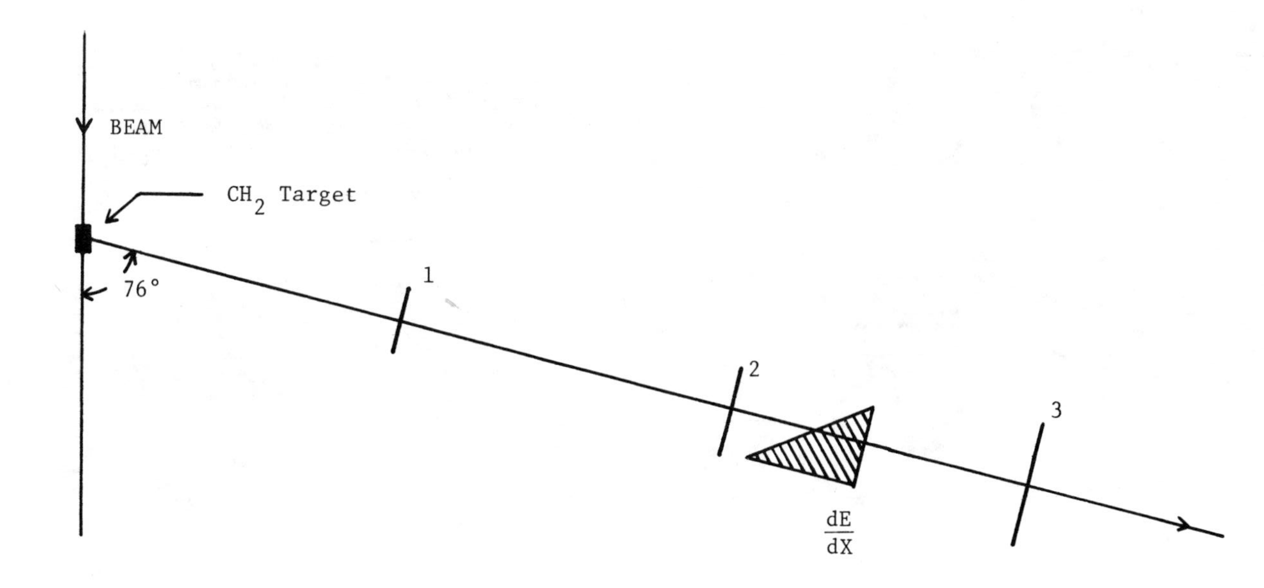

Figure 4. Relative Polarimeter (one of two arms)

Various simple relative polarimeters are then calibrated against this absolute one. A schematic of one of these, the ZGS main ring polarimeter, is shown in Fig.4. There are two identical left and right arms (one of which is shown) each consisting of two scintillator range telescopes. A wedge-shaped absorber is placed in front of the last counter, so that slow recoil protons from elastic and quasi-elastic scatters in the CH_2 with $\langle -t \rangle \sim .15$ stop in that counter. An analyzing power which is about 2/3 A_{pp} is achieved, and with the high event rate the beam polarization can be measured in about 20 sec. with a $\Delta P_B \sim .005$ relative error. Such a rapid measurement is especially useful for tuning through depolarizing resonances in the accelerator cycle. This type of polarimeter has been used by several groups of experimenters.

Polarimeters similar to these might be used after the booster of an accelerator before injection into the main ring or for a proton beam stored in a storage ring for energies up to 25-30 GeV. For higher energy beams, different techniques may have to be used.

VERY HIGH ENERGY POLARIMETERS

A. Inclusive Hadronic Reactions: Large asymmetries have been observed in inclusive $p+p \to \pi^{\pm}+X$ at 6 and 12 Gev/c[3] and $p+p \to \pi^{\circ}+X$ at 24 Gev/c. The $\pi^{\pm}$ asymmetries are large at small transverse momentum ($P_{\perp}$) and large $x=P_{\ell}/P_{max}$ and are energy independent between 6 and 12 GeV/c. These data for x>.7 are plotted in the cross hatched areas in Fig.5. Note that the π^- asymmetry is 30-35% at small u (i.e., small $P_{\perp}$).If this asymmetry remains large at several hundred GeV/c, then large x pion production might be used for a polarimeter. Measurements of these asymmetries can be made using a polarized gas jet or a polarized proton target (PPT) at FERMILAB and the SPS by detecting slow π^- from the target. The experiment is more difficult with a PPT, since its effective polarization is $\sim$7% for this sort of measurement and cannot be rapidly reversed. Nonetheless, if the asymmetries are large a useful measurement might still be made (e.g., $A_{\pi^-} = .35 \pm .07$). The experiment done with a polarized gas jet should be straightforward, and the values of A_{π} should be precise, since

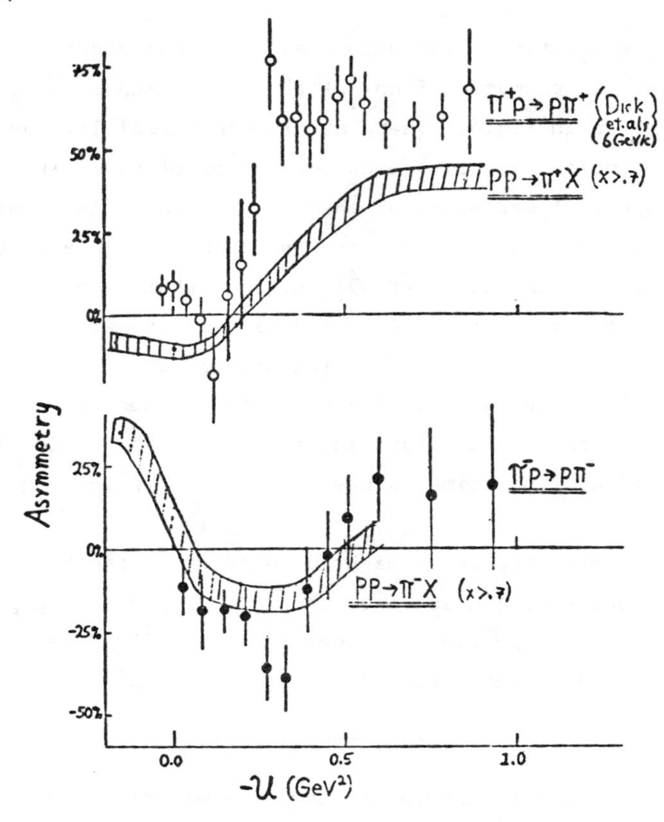

Fig.5. The $\pi^{\pm}$ asymmetry for large x at 11.8 GeV/c compared with the 6 GeV/c polarization data of Dick, et al for backward $\pi^{\pm}p$ elastic scattering.

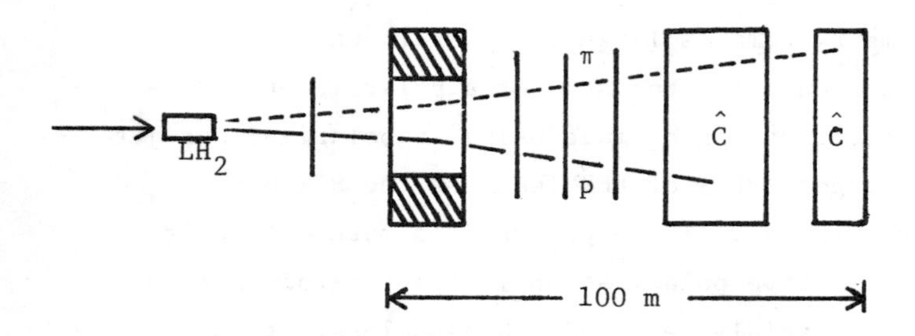

Fig.6. Inclusive Pion Polarimeter

high effective target polarization (∿.80) and rapid spin reversal (∿1 msec) can be achieved. Using the data for large x pion production at 200-400 GeV of Anderson et al.,[5] assuming all π^- for $x>.7$ and $P_\perp^2<.2$ $(GeV/c)^2$ fall within the acceptance of the polarimeter, one gets $50\pi^-$/pulse in scattering a beam of 10^7 protons from a 50 cm liquid hydrogen target. In 100 pulses at FNAL this gives

$$\Delta P_B = \frac{.014}{A_\pi} \cong \pm.05 \quad \text{if } A_\pi = .25-.30$$

This sort of polarimeter would be convenient for a low intensity polarized beam, such as might be gotten from hyperon decays. A spectrometer such as the single arm spectrometer at FNAL could be used as a detector. Detection of π^- might be especially convenient, since π^- bends opposite to the to the beam. Since the asymmetries observed at the ZGS in $pd\rightarrow\pi X$ were the same as in $pp\rightarrow\pi X$, this inclusive polarimeter maybe a good candidate for use with polarized deuteron beams.

Dick, et al.[6] have proposed measuring asymmetries in $p\uparrow p\rightarrow\pi^\circ+X$ by scattering the beam in the main ring of the SPS from a polarized gas jet. Dick et al.[4] have observed large (30%) asymmetries at 24 GeV/c for π° production near x=0 and $P_\perp$∿1. Measurement of asymmetries for π° production in scattering from polarized H and D jet targets could give rise to a convenient design for a "main ring" polarimeter. Since the measurements will be made with a polarized gas jet near x=0, measurements with a dense unpolarized hydrogen gas jet and with an accelerating polarized beam can be made with the same apparatus. Thus there might be a convenient polarimeter installed inside the SPS main ring.

Schematics illustrating the general techniques of apparatus for inclusive pion polarimeters are shown in Fig.6 and Fig.7. In Fig.6 the polarized beam impinges on a liquid H_2 target. The incident beam and the produced π^- are separated by a dipole which provides momentum analysis and sweeping. Proportional chambers or fine scintillator hodoscopes are used for momentum analysis, and the two Cherenkov counters in series cleanly separate pions from protons. Fig.7 shows a possible layout of π° detectors (lead glass Cherenkovs) with appropriate charged particle vetoes and shielding. The π° momentum and

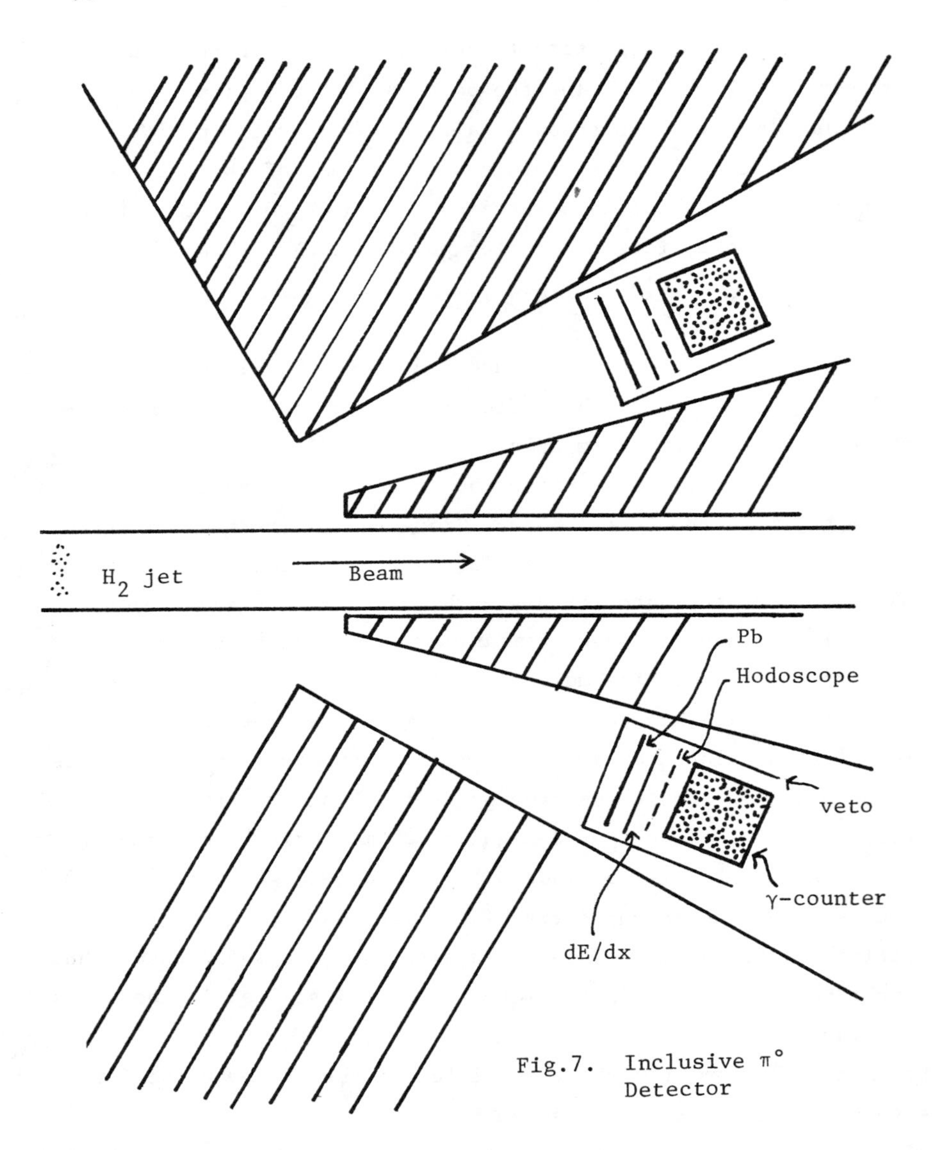

Fig.7. Inclusive π° Detector

thus its $P_{\perp}$, is determined by its energy deposit in the lead glass.

Finally, one might think the reaction $pp \to pX$ would furnish a convenient polarization analyzer, since the cross section integrated for $x=.75$ to $x=.95$ is about 5 mb. However, asymmetries in this reaction measured at 12 GeV[7] are small (0-4%) over the entire kinematic range $x=.1 \to .9$ and $P_{\perp}=0 \to 1.0$ GeV/c. Unless these asymmetries increase in going to higher energies, this reaction will not be useful as a beam spin analyzer.

The inclusive pion production can be used for polarimetry: in an external beam using a liquid H_2 target, inside the main ring of an accelerator using a gas jet target, or for a stored beam in a storage ring using a jet. Dick, Cool et al.[6] have proposed to measure inclusive π° asymmetries using a polarized gas jet inside the main ring of the SPS. The same apparatus could then be used with a very dense unpolarized hydrogen jet as a polarimeter for an accelerating polarized beam in the SPS. The use of a dense hydrogen jet in a storage ring might cause problems with the vacuum in the intersections. Thus a metal vapor jet (e.g., Li), which is easier to collect and pump away might be used instead, since the observed inclusive pion asymmetries are the same, within errors, for pp and pd scattering.

B. Elastic Scattering: for beams of energy $\lesssim 30$ GeV, such as may be stored in the ISR, detecting elastic scattering at small $-t$ (~ 0.1) using a double arm spectrometer will furnish a feasible absolute polarimeter. With polarized deuteron beams of these energies, the quasi-elastic scattering $p(n)p \to pp$ can be detected with a similar spectrometer. A schematic of such a polarimeter is shown in Fig.8. Left (L) and right (R) forward arms detect the scattered polarized beam protons, and the recoil particles from the jet are detected in common recoil detectors on both sides of the intersection. An existing detector, such as the Split Field Magnet, might be used in which elastic scattering events from the jet can be reconstructed to give a measurement of the beam polarization on line. Simple relative polarimeters can be calibrated against this absolute polarimeter as was done at the ZGS. These relative polarimeters can be simple scintillation counter telescopes viewing metal vapor jets located between

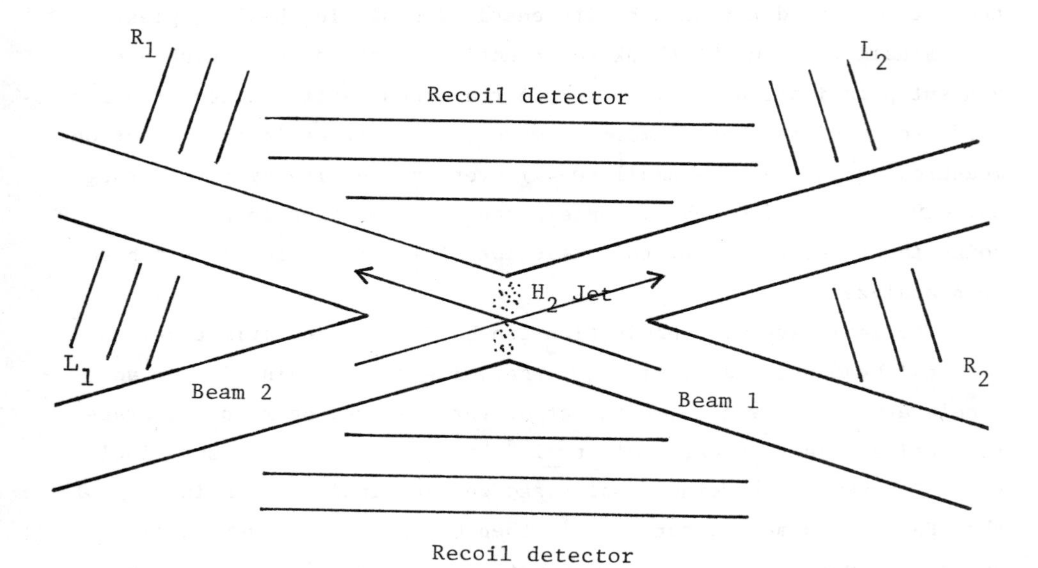

Fig.8. Polarimeter for Intersecting Storage Rings.

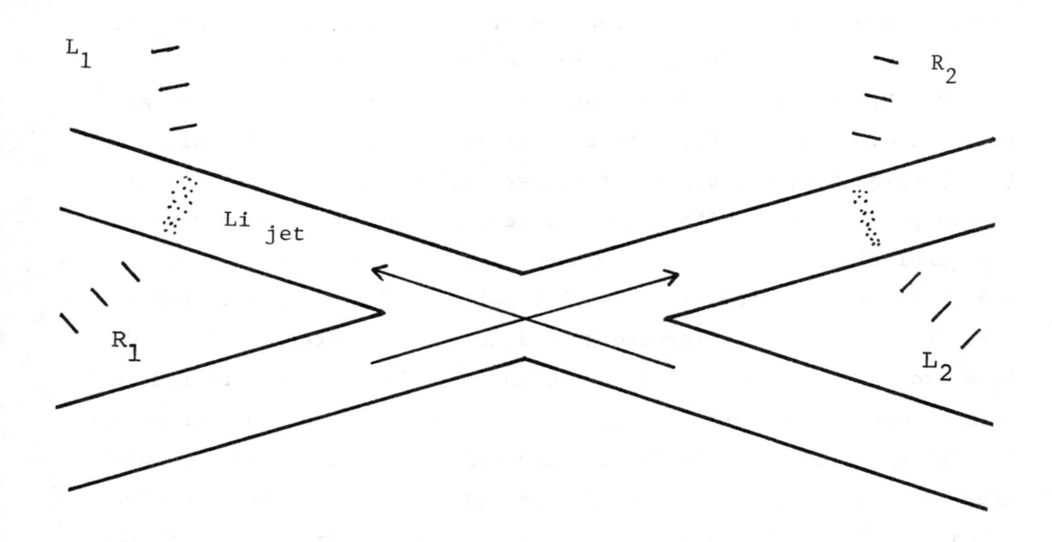

Fig.9. Simple Storage Ring Polarimeter

the intersections to avoid problems with the vacuum. A schematic of such a polarimeter is shown in Fig.9, with left (L) and right (R) telescopes to detect the quasielastic recoil protons from the Li.

Leader et al.[8] and many authors since Schwinger[9] have pointed out that there will be a significant polarization in pp and np scattering in the Coulomb-Nuclear interference region due to the interference of the strong non-flip amplitude and the electromagnetic spin-flip amplitude. For the case of np→np, the polarization of the neutron (which is not equal to that of the proton, since electromagnetic interactions violate isospin invariance) is proportional to $(1+\rho^2)^{-\frac{1}{2}}[\rho=\mathrm{Re}(\phi_1+\phi_3)/\mathrm{Im}(\phi_1+\phi_3)$ at $t=0]$, and should approach 100% where the Coulomb and Nuclear amplitudes are equal $(-t=10^{-5})$. Rosen et al.[10] have proposed construction of a polarized neutron beam using this effect, and Jones, Longo et al.[11] have proposed to measure various spin amplitudes near t=0. This "Schwinger effect" can be used for measuring the polarization of a neutron beam. However, the scattering angles are very small, so very long spectrometers with extremely careful collimation and surveying are needed. The intensity expected with such a beam also limits the experiments that can be done. The advantages are that identical apparatus can be used for polarizer and analyzer and the analyzing power is very large. We shall not discuss this method further, since the main direction of this workshop is toward attaining high intensity polarized proton beams at very high energies. These effects and methods for exploiting them are discussed elsewhere in these proceedings.

The proton polarization calculated by Soffer et al.[12] arising from the Coulomb Nuclear interference in pp scattering is shown in Fig. 10. The polarization is about 5% at $-t=2\times10^{-3}$ and is almost independent of energy. This calculation assumes the strong spin flip amplitude $\phi_5=0$ for very small -t; Dick, Cool et al. have proposed to measure the pp asymmetry (equal to the polarization by time reversal invariance) in this t range, so even if Im ϕ_5 is not neglible, measured A_{pp} values will be available. The fact that the energy independent Coulomb induced polarization is ∿5% and has the same sign as the hadronic polarization indicates that this may possibly be the

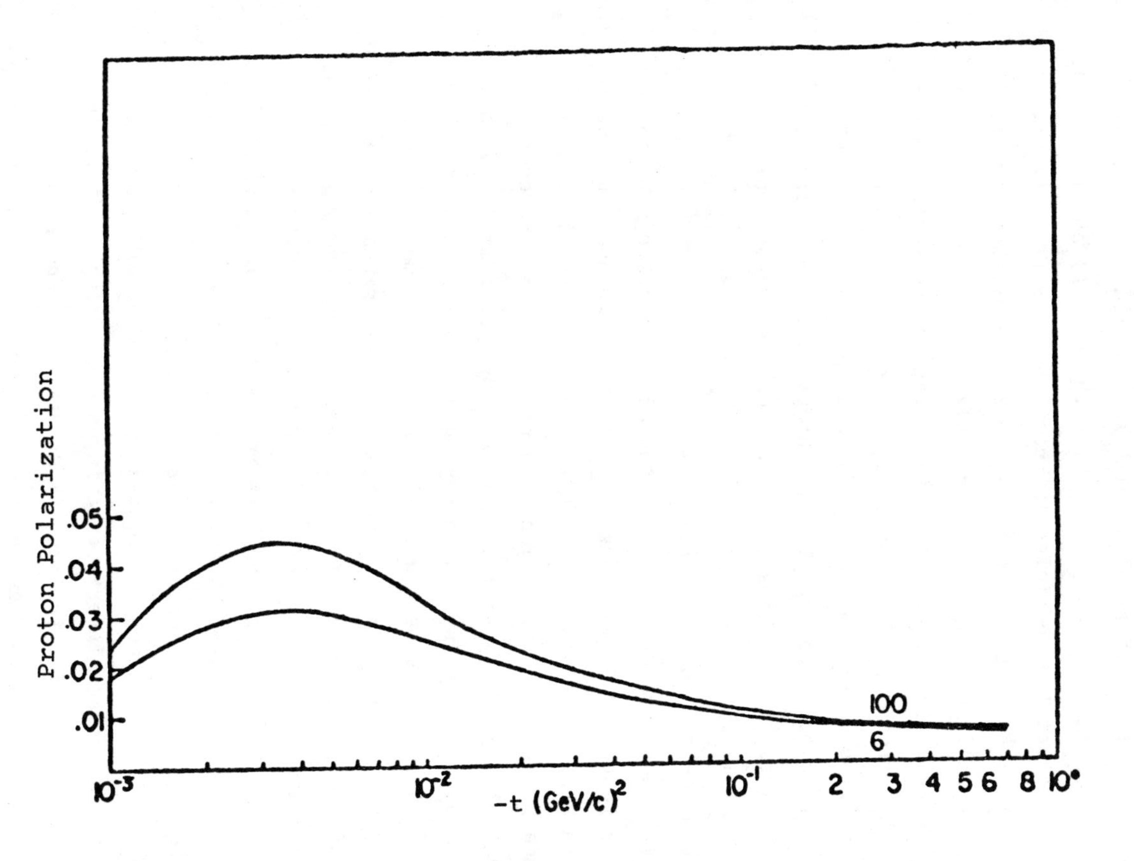

Fig.10. Proton Proton Polarization from Coulomb-Nuclear Interference

best mechanism for construction of a polarimeter at very high energies, since the rates are quite high in the Coulomb region.

The proposed apparatus of Dick, Cool et al. could be used as a polarimeter for an accelerating polarized proton beam. A schematic of this apparatus is shown in Fig.11. The array of solid state detectors covers $\Delta\phi=\pm45°$ from the median plane. If the useful $-t$ range is $.001<-t<.01$ for an average analyzing power of $\sim5\%$, and a dense hydrogen jet can give a luminosity of $3\text{x}10^{33}$/sec interacting with the circulating accelerated beam, there are $\sim2\text{x}10^{3}$ events/pulse useful for beam polarization measurement. This will give an error $\Delta P_B=\pm0.1$ in about two minutes at the SPS. The time can be shortened and the accuracy of measurement can be improved by a factor of 2-3 by using larger solid state counters. The background rates due to upstream beam gas interactions, beam loss, etc., may have to be suppressed by veto counters. Hopefully such an apparatus will be made to work by the above group before it is needed as a polarimeter. This configuration is not useful as a polarimeter for deuterons, since the Coulomb induced polarization is proportional to the anamolous magnetic moment, which is too small to yield useful analyzing power for dp scattering. The method may also be useful for polarized beams derived from hyperon decay. With the above Δt and $\Delta\phi$, in scattering a beam of 10^7 polarized protons from a 20" liquid H_2 target, there are $8\text{x}10^4$ events/20 mintues at FNAL or SPS, which gives $\Delta P_B=\pm0.1$ for an analyzing power of 4%. The Θ_{rms} for multiple Coulomb scattering in the liquid H_2 target is an order of magnitude less than the single Coulomb scattering angles, and thus does not jeopardize the measurement.

We thus conclude that elastic scattering in the Coulomb-Nuclear interference region should prove to be a very useful analyzer for the polarization of very high energy beams of protons.

C. Diffractive Processes: strong diffractive processes, such as diagrammed in Fig.12, could show strong spin effects at very high energies, since the relative energy for the πp scattering is low. However, there are strong final state interactions, and interpreting the spin effects in terms of low energy πN polarization is quite

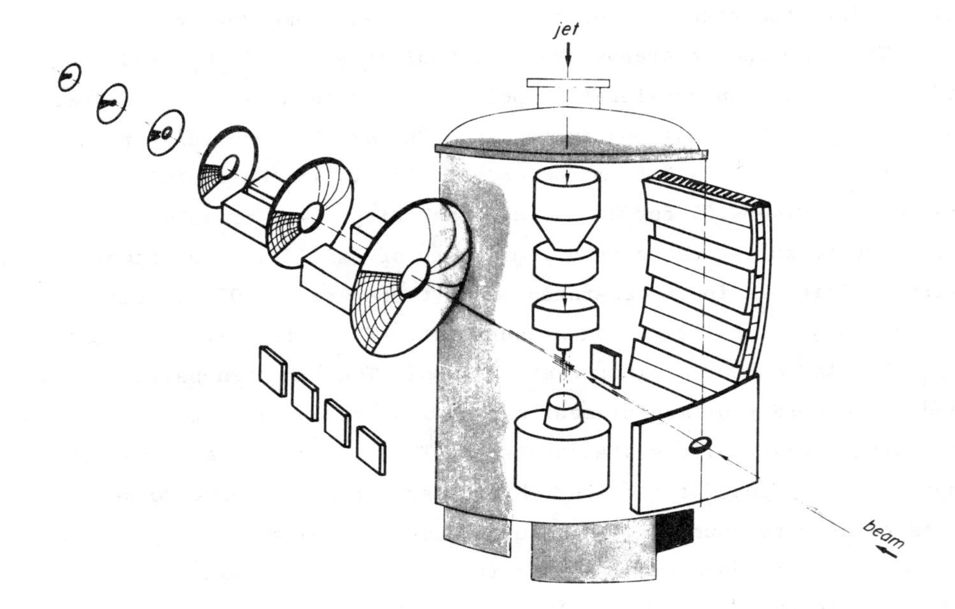

Fig. 11

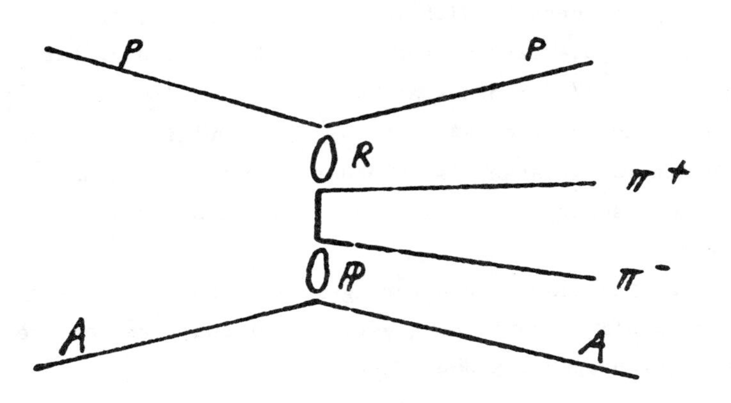

Fig.12.

ambiguous. Even if polarization effects in this sort of interaction could be directly measured at very high energy, untangling the effects of the initial beam spin and the final state decay would require a complicated reconstruction and analysis of final particle distributions.

A physically simpler process which can be related to low energy phenomena is the Coulomb diffractive dissociation of protons into the $N\pi$ system. In particular, the process $pA \rightarrow pA\pi^\circ$ can be related via the Primakoff effect to low energy photoproduction, i.e., $\gamma p \rightarrow \pi^\circ p$. In the diagram below the amplitude ψ for diffractive production from a nucleus with charge Z and atomic number A is

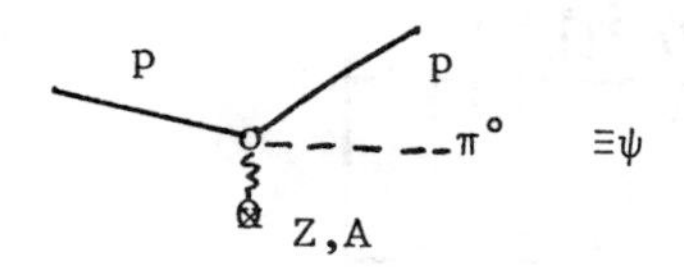

At this workshop, G. Thomas has shown explicitly that

$$\int |\psi|^2 d\phi = \frac{Z^2 \alpha}{P_\perp^2} \left(\frac{2M_A}{s_{\pi p} - m_p^2} \right)^2 \left[A_\gamma \frac{d\sigma}{d\Omega} (\gamma p\uparrow \rightarrow \pi^\circ p) \right] \qquad (1)$$

where A_γ is the photoproduction asymmetry at the given value of $P_\perp$ and $s_{\pi p}$.

The γp effective kinetic energy is typically 500 MeV, yielding asymmetries $\sim$40% as shown in Fig.13. The $p\pi^\circ$ effective mass should probably be above the Δ, since asymmetries on the Δ mass may be small. Measurements of the photoproduction asymmetry at smaller missing masses than for present data are clearly necessary. Integration over ϕ, the angle between the N^* production plane and the $p\pi^\circ$ decay plane is clearly necessary, as can be seen from equation (1), for a clean measurement of the beam polarization. As an aside, since the virtual γ in the above diagram is longitudinally polarized, the electromagnetic spin correlation allows one to analyze the longitudinal polarization of the incoming proton, if the correlation in $\vec{\gamma}\vec{p} \rightarrow \pi^\circ p$ is known at the appropriate low energy. A possible apparatus for effecting this sort of polarimeter is shown in Fig. 14. The final state $\pi^\circ p$ must be detected with very good angular resolution, since the low energy

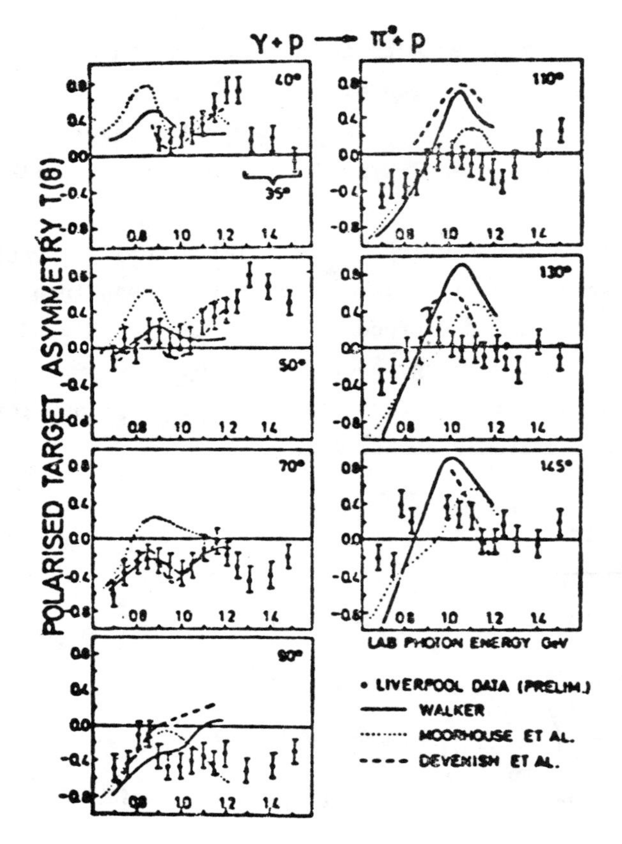

Fig.13. Polarized target asymmetry in pion photoproduction.

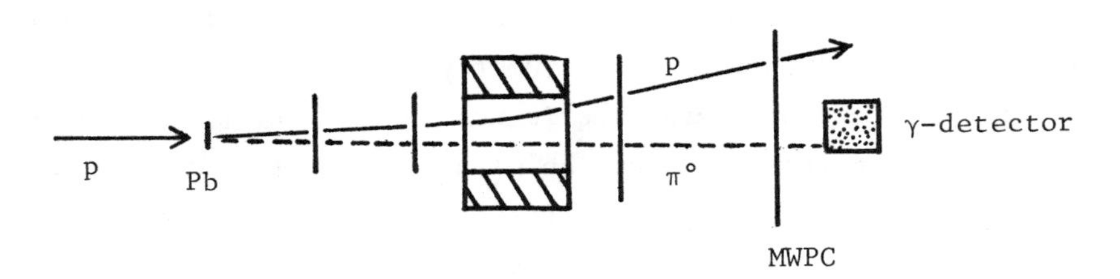

Fig.14. Polarimeter using Coulomb diffractive production

photoproduction asymmetry may vary rapidly with c.m. angle. The total cross section for the Coulomb dissociation is very large (∿20mb), so the rates with the shown apparatus are quite adequate for a low intensity polarized beam such as might be derived from hyperon decays. (For a high intensity beam going through the apparatus, or a beam internal to a main ring, or a storage ring, the process is probably not appropriate.) Since scattering from a heavy nuclear target gives a large gain in rate, we use a 3 mm Pb target which outscatters about 10^{-5} of the beam. For 10^7 incident polarized protons with the apparatus shown one would get about 100 events/FNAL burst in the region of useful asymmetry for $M_{p\pi^\circ} \sim 1300$ Mev. Thus a measurement of the beam polarization with $\Delta P_B \cong \pm .03$ can be made in about 20 minutes at FNAL or the SPS. However, this requires a measurement of the ($p\pi^\circ$) angle to .03 mR so that angular resolution of ∿5° in the equivalent low energy γp system can be achieved. Since the asymmetry can vary rapidly, this c.m. angular resolution is essential. Measurements must be made for $|t'|<.001(\mathrm{GeV}/c)^2$ to cleanly separate the electromagnetic production from the strong diffractive processes. The polarimeter would thus look similar to the apparatus for FNAL E272,[14] and the on line reconstruction of beam polarization would be difficult. However, for polarized proton beams from hyperon decays, this measurement time should be adequate.

CONCLUSIONS

We conclude that there are several feasible designs for polarimeters for high energy beams of polarized protons. In particular, the use of Coulomb scattering seems to be very promising. Possibilities for deuteron polarimeters are more limited. Measurement of various elastic and inelastic spin effects at SPS and FNAL energies using polarized proton targets and polarized hydrogen and deuterium gas jets is strongly encouraged. In addition to the possibility of making exciting physics discoveries, better spin analyzers for polarized proton and deuteron beams may be found.

REFERENCES

[1] High Energy Physics with Polarized Beams and Targets AIP Conf. Proc. No. 35, 1976 (M. L. Marshak, editor) pp. 145,152. Results were also presented at this workshop by O. Chamberlain and G. Fidecaro, and appear elsewhere in these proceedings.

[2] Figure 2. is from L. G. Ratner *et al*., Phys.Rev. D15 (1977) 604. Figure 3. is from R. D. Klem *et al*. Phys.Rev. D15 (1977) 602.

[3] R. D. Klem *et al*., Phys. Rev. Lett. 36 (1976) 929.

[4] J. Antille *et al*. to be published, privately communicated by L.Dick at this workshop.

[5] R. L. Anderson *et al*., Phys.Rev.Lett. 37 (1976) 1111.

[6] Study of Spin Effects in pp Reactions at SPS Using a Polarized Atomic Hydrogen Gas Jet. SPS Proposal SPSC/88, 26 August 1977.

[7] AIP Conference Proceeding No. 35 (M.L. Marshak, editor) p. 219 and ff., and submitted to The Physical Review.

[8] E. Leader, presentation at this workshop. Work done at the workshop and earlier work done in conjunction with N. Buttimore and E. Gotsman are published elsewhere in these proceedings.

[9] J. Schwinger, Phys. Rev. 73 (1948) 407.

[10] J. Rosen, Proceedings of the Summer Study on High Energy Physics with Polarized Beams, Argonne 1974; ANL preprint ANL/HEP 75-02 (J.B. Roberts, editor).

[11] L. W. Jones, et al. ZGS letter of intent LI-71.

[12] C. Bourrely and J. Soffer, Nuovo Cimento Lett. 19, 569 (1977).

[13] Use of this reaction for a polarimeter was first suggested by D. Underwood and will be extensively discussed in ANL preprints now in preparation.

[14] D. Underwood, private communication.

POLARIZED PROTON BEAMS PRODUCED BY HYPERON DECAY : SUMMARY OF DISCUSSIONS

G. Fidecaro
CERN - European Organization for Nuclear Research,
1211 Genève 23, Switzerland

INTRODUCTION

The review talk on polarized proton beams produced by hyperon decay at very high energy, and the discussion which took place in connection with this talk, indicated that the following main topics deserved further consideration:

a) problems of beam design,
b) measurement of the polarization of the beam,
c) feasibility of some experiments.

PROBLEMS OF BEAM DESIGN

It was estimated that specific problems of beam design could perhaps be better examined in the light of a comparison between the two projects under consideration at CERN and at FNAL.

A comparison between these two designs is shown in the Table reproduced below. The CERN project had already been optimized at the engineering level and detailed information was therefore easily available (see the Review Talk). Nevertheless, some figures were updated after the Worskhop, thanks to the work of N. Doble [1]. The FNAL beam was still under study and the relevant information was given to the Workshop by A. Yokosawa or taken from the report presented by David Underwood at the 1977 Argonne Symposium on Experiments using enriched antiproton, polarized proton and polarized antiproton beams.

Going through the Table one can notice a difference in the intensity of the primary proton beam. However, this difference should not be regarded as entirely meaningful. In fact, also at CERN it would be possible to have 10^{13} primary protons per pulse on a single target; however, in this case, the running time would have to be necessarily shorter because a higher intensity would imply shifting the competition with the other users of the accelerator from intensity to machine time, so that in the end only the integrated primary beam intensity would still matter.

The same Λ production spectrum [2] was used in both cases.

The length of the primary target had been optimized at CERN, in order to take into account the hyperon absorption. At FNAL, the length of the target was that still currently used today.

ISSN: 0094-243X/78/087/$1.50

TABLE

Comparison of the "CERN" and "FNAL" projects
for a polarized proton beam produced by hyperon decay

	CERN	FNAL
Energy of primary proton beam	400 GeV	400 GeV
Intensity of primary proton beam	3×10^{12} ppp	10^{13} ppp
Target	40 cm Be	20 cm Be
Decay length	16.5 m	$\sim$ 20 m
Acceptance	1.6 μsr	$\sim$ CERN's x 2
Momentum analysis	in vertical plane	in vertical plane
Momentum band $\Delta p/p$	±1.6%	±2 - ±5%
Horizontal deflection	corresponding to 360° spin rotation at 325 GeV	none
Vertical deflection	compensated	compensated
Horizontal beam size (r.m.s.)	±2.4 mm	not known
Vertical beam size (r.m.s.)	±4.2 mm	not known
Optimum energy of polarized protons	325 GeV	340 GeV
Optimum polarization	35% (longitudinal)	40% (longitudinal)
Intensity of polarized protons	2×10^6 pppp	10^8pppp
Rotation of the polarization vector	by changing direction of beam	change of direction of beam compensated
Calibration	Coulomb-Nuclear interference	Primakoff effect
Physics motivation	Differential elastic scattering on polarized protons σ_{tot} Coulomb-Nuclear interference	Inclusive π^- production up to $p_T \sim 6$ GeV/c σ_{tot}

The final result of these differences, and of the others shown in the Table (larger acceptance through the use of superconducting quadrupoles, and a larger momentum band), is an intensity 2×10^6 polarized protons per pulse at CERN, against 10^8 pppp at FNAL. The discrepancy is rather large; however, as it appears difficult to explain it with the optimization of acceptance and decay length of the hyperons, and with the optimization of momentum band and transmission of the beam transport system, one will have to wait for the final FNAL design in order to understand where a factor as large as 50 can possibly come from.

The possibility of changing the direction of the polarization vector in order to measure up-down or forward-backward asymmetries with a longitudinally polarized proton beam was considered with attention both at CERN and at FNAL. At CERN, a system of two magnets placed just before the experimental target deflects the beam first in one direction and then in the other direction, in the horizontal plane, in such a way to bring the beam to the same place (experimental target), but at an angle with the undeflected beam line. The relative direction of the polarization vector is determined by this angle. The CERN design makes use essentially of the fact that a small change in the direction of the beam makes the polarization vector rotate by a large angle (at 300 GeV a deflection of 2.7 mrad is sufficient to rotate the polarization vector by 90°). This method has the inconvenience that the angle the beam makes with the symmetry line of the experimental apparatus changes when the polarization is reversed. However, this fact was not considered a major difficulty by the experimental groups interested in this polarized beam, since they are accustomed to such a device at present used to steer the beam to the polarized proton target, in order to compensate the deflection produced by the polarized target magnet.

The FNAL design makes use of a set of bending magnets working some in the horizontal plane, others in the vertical plane, in order to rotate the spin without changing the direction of the beam. The basic idea is to decouple the direction of the spin from the direction of the beam by working on both planes : the transversally polarized proton beam obtained by a first deflection, for example in the horizontal plane, of the original longitudinally polarized proton beam, is made to go through a second bending magnet working in the vertical plane , whose field is therefore parallel to the spin. This magnet deflects the beam down in the vertical plane by an angle corresponding to a rotation of the spin by 90°, without affecting the spin. A third horizontal bending magnet then re-aligns the spin on the momentum of the beam (not yet horizontal), and a fourth vertical magnet while making the longitudinally polarized beam horizontal again, by bending it up, converts it into an "up" transversally polarized proton beam. By reversing the polarity of all magnets one would obtain a "down" transversally polarized beam. However, though the beam maintains its original direction while going through the four-magnet system described above,it is displaced in the horizontal and in the vertical planes, with respect to its original line. Obviously, when the polarization is reversed, the beam is displaced in the opposite direction. These displacements can be compensated by an additional set of magnets which always brings the beam back to its original line. A similar technique can be used to reverse the polarization vector of a longitudinally polarized proton beam. The scheme is essentially that invented by Derbenev, Kondratenko and Skrinsky at Novosibirsk (see the Report of the Working Group on Accelerator Problems : "The Siberian Snake"). A similar method was used on the E61 experiment at FNAL to eliminate asymmetries from any unwanted polarization of the secondary proton beam.

Just before this Worskhop began, the CERN groups were also considering adopting the same method. They plan to compensate for the simultaneous displacement of the beam in the vertical and in the horizontal planes by having the four bending magnets mentioned above preceded by two additional bending magnets working in a plane inclined at 45°. The forward-backward rotation of the spin can be obtained by an additional magnet placed at the end of the beam, rotating the spin by 90° always in the same direction. This last magnet would convert, for example, an "up" polarization into a forward one and a "down" polarization into a backward polarization [3].

The physics motivation which led the authors of the two proposals to become interested in polarized proton beams are also different, with some overlap. In both cases the main experimental effort (differential elastic cross sections on a polarized target at CERN; inclusive π^- production up to p_T = 6 GeV/c at FNAL) would require a rather large intensity, or else the t or p_T range would be correspondingly decreased. It is worth remembering that the CERN groups had started working on their experimental proposal while relying on a polarized beam with an intensity a good order of magnitude larger and with larger polarization [4]. The more accurate calculations made when the new production spectra became available [5] made the intensity go down. This indicates that the intensity of polarized proton beams from hyperons depends critically on the shape of the production spectrum, thus proving the necessity for direct measurements to be done at least on a test beam.

MEASUREMENT OF THE POLARIZATION OF THE BEAM

As shown in the Table, two different methods were proposed at FNAL and at CERN for the absolute measurement of the polarization of a polarized proton beam produced by hyperon decay : while the Primakoff effect was examined at FNAL, the CERN groups fixed their attention on the Coulomb-Nuclear interference.

Both methods raised considerable interest among the participants at the Workshop and were therefore discussed at length : in the theory sessions, because participants wanted the margin of uncertainty of the theoretical predictions to be thoroughly discussed, and obviously in the polarimeter sessions, though no direct comparison of the two methods was made.

The method based on the Primakoff effect (see the report on the polarimeter sessions) was estimated to be adequate for the measurement of the polarization of polarized proton beams produced by hyperon decay as it could give the requested information in a time as short as 20 minutes.

The CERN proposal to use the Coulomb-Nuclear intereference, on the contrary, did not aim at a high counting rate because it was directed mainly towards the execution of single precise experimental determination with a view to obtaining information of interest for physics, and less towards the development of a tool for quick repetitive measurements.

However, G. Fidecaro pointed out during the Workshop that there were no reaons why the full intensity of the polarized proton beam could not be used for the measurement of the polarization. With an intensity of 2-3 x 10^6 pppp and a t-band of $\sim$ 0.01 GeV^2, the time to get a statistical accuracy of ±0.5% on the polarization could in fact be decreased from 4 days down to 15 or 20 minutes, and the device used as a polarimeter. This could be obtained with a set-up substantially similar to that proposed by CERN, equipped with a fast electronic computer for the determination of the scattering angle [6].

FEASIBILITY OF SOME EXPERIMENTS

In the course of the discussion of the specific experiments proposed for the two polarized proton beams by hyperon decay here under examination, the expected statistical errors computed at CERN for the measurement of the correlation parameter A_{NN} were considered in some detail. In fact, the feasibility of this type of experiments - given the fairly low intensity expected at CERN - depends critically on these errors, and to some of the participants the CERN errors looked rather low, and therefore promising, to the point that a confirmation of the latter would have been welcome!

In the discussion which followed, since some of the participants had used the E61 apparatus as reference, a comparison between this experiment and CERN's WA6 became unavoidable. The CERN errors appeared to be substantially correct, when it was recognized that the double-arm WA6 spectrometer proposed for the measurement of the correlation parameter A_{NN} had a counting rate 8 times higher, for the same beam intensity, that that of the corresponding E61 apparatus. A factor of two came from the length of the polarized target which is twice as long at CERN, and a second factor of two from the addition of a second arm to make the WA6 apparatus symmetrical. A third factor of two came from the larger acceptance of the recoil telescope owing to the use of the polarized target magnet itself as analysing magnet for the recoil particle.

For the experiments which could not be done with polarized proton beams in general, see the report of the experimental sessions.

CONCLUSIONS

One could say that a polarized proton beam produced by hyperon decay could hardly replace an accelerated beam to carry out a complete programme of measurements, because of the small intensity and low polarization obtainable. However, it would be suitable for exploratory work and for specific measurements, while waiting for progress in the field of accelerators, especially if the ideas at present under discussion at FNAL in connection with the design of their polarized proton beam from hyperon decay should turn out to be correct - and this beam is able to furnish an intensity substantially higher than that foreseen at CERN - and if more efficient apparatus could be built, as indicated by the CERN groups.

REFERENCES

1. N. Doble (CERN), private communication.
2. P.L. Skubic, University of Michigan - Thesis - UM HE 77-32.
3. Supplement to the Proposal SPSC/P 87 - SPSC/M 94 - Oct. 21, 1977.
4. Annecy, CERN, Marseilles, Padua, Serpukhov, Trieste, Vienna - On the study of spin effects at SPS energies by making use of a polarized beam facility - Letter of intent SPSC/I 87 - 11 Nov. 76.
5. G. Bunce, K. Heller and O. Overseth, private communication.
6. A tentative design of a fast scattering angle computer was made at CERN after this Workshop : G. Fidecaro and I. Pizer, private communication.

REVIEWS

WORKSHOP ON POLARIZED PROTON BEAMS
ANN ARBOR, MICHIGAN, OCTOBER 18-27, 1977

ACCELERATION AND STORAGE OF POLARIZED BEAMS

E. D. Courant*
BNL and SUNY, Stony Brook

INTRODUCTION

I. Review of Theory

The polarization of a beam in a circular accelerator or storage ring is affected by depolarizing resonances. The theory of these resonances is well known;[1] however, I shall formulate it here in such a way as to facilitate quantitative estimates, in terms of the detailed characteristics of a particular accelerator.

The Froissart-Stora (FS) equation is:

$$\frac{d\vec{S}}{dt} = \frac{e}{\gamma mc} \vec{S} \times [(1 + \gamma G) \vec{B}_{\perp} + (1 + G) \vec{B}_{\parallel}] \tag{1}$$

where $G = (g - 2)/2$ in the gyromagnetic anomaly, and $\vec{B}_{\perp}$ and $\vec{B}_{\parallel}$ are the parts of the magnetic field perpendicular and parallel to the direction of motion.

It is useful to rewrite the FS equation in terms of the motion of the particle in the field of the accelerator. We assume the accelerator or storage ring has a reference orbit lying in the horizontal plane (not necessarily the equilibrium orbit of a particle - field errors may shift the particle orbit from the reference orbit). Define the following local coordinate system.[2]

*Work performed under the auspices of the U.S. Department of Energy.

1. M. Froissart and R. Stora, Nucl. Instrum. 7, 297 (1960).
2. See, for example, E.D. Courant and H.S. Snyder, Ann. Phys. 3, 1 (1958).

ISSN: 0094-243X/78/094/$1.50

s = distance along reference orbit

$\rho(s)$ = radius of curvature

$\vec{b}$ = unit vector along reference orbit

$\vec{a}$ = unit vector in orbit plane transverse to orbit

$\vec{c}$ = $\vec{a} \times \vec{b}$ = vertical unit vector

The equation of motion of the particle:

$$\frac{d\vec{v}}{dt} = \frac{e}{\gamma mc} \vec{v} \times \vec{B} \tag{2}$$

then enables us to write $\vec{B}_{\perp}$ and $\vec{B}_{\parallel}$ in terms of the radius of curvature and the coordinates of motion of the particle (including the deviations from the reference orbit due to field errors as well as betatron oscillations):

$$\frac{d\vec{S}}{ds} = \vec{S} \times \vec{F}$$

with

$$\vec{F} = \left[(1 + G)\left(\frac{z}{\rho}\right)' - (1 + \gamma G)\,\frac{z'}{\rho}\right]\vec{b} + (1 + \gamma G)\left[z''\,\vec{a} + \frac{\vec{c}}{\rho}\right] \tag{3}$$

where z is the vertical excursion of the particle and the prime denotes differentiation by s.

Now describe the spin vector $\vec{S}$ in terms of its components S_1, S_2, S_3, in the $\vec{a}$, $\vec{b}$, $\vec{c}$ directions; introduce as the new independent variable the turning angle $\theta = \int ds/\rho$ and formulate the spin motion in terms of 2-component spinors rather than 3-component vectors (as first done by Hamilton a century ago). With the wave function ψ, we have:

$$S_i = \psi^{+} \sigma_i\, \psi, \quad i = 1, 2, 3$$

where σ_i are the Pauli spin matrices, and we find:

$$\frac{d\psi}{d\theta} = \frac{i}{2} \begin{pmatrix} -\varkappa & \zeta \\ \zeta^* & +\varkappa \end{pmatrix} \psi \tag{4}$$

where

$$\varkappa = \gamma G \tag{5}$$

and

$$\begin{aligned} \zeta\,(s) &= -\rho \left(F_a - i\, F_b\right) \\ &= -\,(1 + \varkappa)\,(\rho z'' + i z') + i\rho\,(1 + G)\left(\frac{z}{\rho}\right)' \end{aligned} \tag{6}$$

For a particle on the reference orbit, $\zeta = 0$, and σ_3 is constant, while σ_1 and σ_2 precess with frequency $\varkappa$. But if ζ possesses a component oscillating as $e^{-i\varkappa\theta}$, a resonance will occur.

The frequencies contained in ζ are determined by the behavior of $z(\theta)$. Accelerator orbit theory[2] tells us that, in general

$$z(\theta) = z_e(\theta) + \tfrac{1}{2}\sqrt{\varepsilon\,\beta(\theta)}\left[e^{-i[\nu\theta - \chi(\theta)]} + c.c.\right] \quad (7)$$

where $z_e(\theta)$ is the closed orbit whose deviation from zero is produced by field errors, and the second term is the betatron oscillation; here ν is the "tune" or oscillation frequency, $\beta(\theta)$ and $\chi(\theta)$ are certain periodic functions of θ having the azimuthal periodicity of the accelerator structure. The radius of curvature ρ has this same periodicity. ε is the emittance.

Therefore, components with frequency $\varkappa$, exist in ζ if we have:

$$\varkappa = kP \pm \nu \quad (8)$$

for intrinsic resonances excited by betatron oscillations, where P is the number of identical periods in the accelerator structure and k is any integer. Imperfection resonances arise at all integral values of $\varkappa$, since the closed orbit contains all harmonics; the most prominent ones are again likely to be the ones near the structure resonances.

II. EFFECT OF RESONANCE

a. Steady State

If the fields are time-independent, we suppose that we are a distance δ from a resonant value of $\varkappa$, and that ζ contains a resonant component. Then (4) becomes:

$$\frac{d\psi}{d\theta} = -\frac{i}{2}\begin{bmatrix} \varkappa_o + \delta & \epsilon e^{-i\varkappa_o\theta} \\ \epsilon e^{i\varkappa_o\theta} & -(\varkappa_o + \delta) \end{bmatrix}\psi \quad (9)$$

This is easily solved; in general, the polarization,

$$P = S_3 = \psi^+ \sigma_3 \psi$$

oscillates rapidly around a mean value. The maximum time-average or steady-state polarization can be seen to be:

$$P_{max} = \frac{\delta}{\sqrt{\epsilon\epsilon^* + \delta^2}} \qquad (10)$$

Therefore, $|\epsilon|$ may be called the width of the resonance.

b. Passage through Resonance

Now if one passes through a resonance by changing δ with time from negative to positive, this equilibrium spin will change sign. This is just the phenomenon of "fast adiabatic passage" familiar in the field of magnetic resonance in solid-state physics (the term "fast" refers to relaxation times which are irrelevant in our context).

In general, passage through resonance is analyzed by putting $\delta = \alpha\theta$ in (9), where:

$$\alpha = \frac{d\varkappa}{d\theta} = \frac{1}{2\pi} \quad (G\ \Delta\gamma \pm \Delta\nu) \qquad (11)$$

where $\Delta\gamma$ is the energy gain per turn and, for betatron resonance, $\Delta\nu$ is the tune shift per turn due to pulsed quadrupoles such as those used at Argonne. [For imperfection resonances the $\Delta\nu$ term in (11) is omitted.]

The solution, first given by Froissart and Stora,[1] is obtained by transforming (9) to a confluent hypergeometric equation. It is found that, with initial polarization P_1, the polarization after passage through resonance is:

$$P_2 = P_1 \ (2e^{-\pi|\epsilon|^2/2\alpha} - 1) \qquad (12)$$

therefore, polarization remains to the extent of 99% if,

$$x = \frac{\pi}{2} \ \frac{|\epsilon|^2}{\alpha} < 0.005, \text{ or } |\epsilon| < 0.03 \sqrt{\Delta\gamma \pm \Delta\nu/G}$$

(for protons with G = 1.79).

Spin-flip is 99% complete, i.e., polarization is present with change of sign, if,

$$|\ x\ | > \ell n\ 200 = 5.30,$$

$$|\epsilon| \quad > 0.980 \sqrt{\Delta\gamma \pm \Delta\nu/G}.$$

In practice, it is found at Argonne that complete spin-flip is difficult to obtain, probably because synchrotron oscillations ensure that not all particles have the same $\Delta\gamma$, and also because some particles have a small emittance and, therefore, a small value of ϵ.

III. EVALUATION OF RESONANCE WIDTH

The strength ϵ of a resonance is obtained by finding the component oscillating as $e^{-i\varkappa\theta}$ in ζ as given by (6):

$$\epsilon = \frac{1}{2\pi} \int_0^{2\pi} \zeta \, e^{i\varkappa\theta} \, d\theta \tag{13}$$

In a lattice made up of bending magnets (with or without gradiants and edge focusing) and quadrupoles, one can evaluate (13) analytically, using (7) and the equations of motion satisfied by $z(\theta)$. It is found that, for the structure resonance $\varkappa=\gamma G=kP \pm \nu$, the contribution of a magnet with curvature $1/\rho$, gradient $K\,B\rho$ (K positive for vertical focusing), and entrance and exit wedge angles ζ_1 and ζ_2 is:

$$\epsilon = \frac{\sqrt{\mathcal{E}}}{4\pi} \left\{ \left[C + \frac{(1+\varkappa)}{\rho} (\zeta_1 + i) \right] \beta_1^{\frac{1}{2}} e^{i\chi_1} + \left[C^* + \frac{(1+\varkappa)}{\rho} (\zeta_2 - i) \right] \beta_2^{\frac{1}{2}} e^{i\chi_2} \right\} \tag{14}$$

where,

$$C = \frac{[(1+\varkappa)\;K\rho^2 - \varkappa(\varkappa - G)]}{\varkappa^2 - K\rho^2} \left[\frac{\sqrt{K}\,(\cos \ell\sqrt{K} - e^{i\varkappa\theta})}{\sin \ell\sqrt{K}} + \frac{i\varkappa}{\rho} \right] \tag{15}$$

where ℓ = length of magnet

$\theta = \ell/\rho$ = turning angle of magnet.

The amplitude function β and the phase function $\chi = \nu\theta - \psi$ (ψ = betatron oscillation phase) may be obtained at the beginning and end of each magnet using computer programs such as SYNCH or AGS.

In the <u>smooth</u> approximation for a perfect machine, only the resonance $\varkappa = \nu$ comes into play; in that case $K\rho^2 = \nu$, $\beta = R/\nu$,

$\chi = 0$, $\theta = 2\pi$ and (15) becomes, for the complete ring,

$$\epsilon = \tfrac{1}{2} \sqrt{\frac{\mathcal{E}}{R}}\, \varkappa^{\frac{1}{2}} (\varkappa^2 + G). \tag{16}$$

In the case of imperfection resonances (at integral values of $\varkappa$) one obtains, for the contribution of a magnet,

$$\epsilon = \frac{1}{2\pi} \left\{ \left[C + \frac{(1+\varkappa)}{\rho} (\zeta_1 + i) \right] z_1 + \left[C^* + \frac{(1+\varkappa)}{\rho} (\zeta_2 - i) \right] z_2 \right\} \tag{17}$$

where z_1 and z_2 are the displacements of the closed orbit at the entrance and exit of the magnet.

IV. POSSIBILITIES OF ACCELERATING POLARIZED PARTICLES

The computer program SYNCH or AGS (or equivalent) can evaluate the parameter needed for (14) and (17) for each magnet, and another program DEPOL has been written to evaluate (14) and (17) using the output of SYNCH. For the Brookhaven AGS, with magnets randomly misaligned by ± 0.1 mm and with a beam emittance of $(10\,\pi/\gamma) \times 10^{-6}$ meter-radians, the results are shown in Fig. 1. Here the two horizontal dashed lines correspond to a depolarization of less than 1% per resonance (lower line) and 99% spin reversal (upper line.)

It is seen that below about 20 GeV, there are six intrinsic resonances and about eight imperfection resonances, strong enough to depolarize the beam by more than 1%.

Comparable resonances have been overcome at the Argonne ZGS by fast pulsing of the tune and careful adjustment of magnet orbit errors. It might be feasible to do the same at the AGS (or CERN PS or similar machines) at energies up to about 20-25 GeV.

A complementary method may be to enhance resonances - slow passage through them - until they become strong enough for complete spin-flip. Argonne's experience shows that this may be difficult, probably because synchrotron oscillations prevent all particles from passing through a resonance slowly.

It thus appears possible - but not easy - to maintain proton polarization in the AGS up to about 20 GeV.

An alternative approach is to consider deuterons. The deuteron anomalous moment G, is only -0.14 and its mass is twice that of the proton; therefore, the resonances are spaced 25 times as far apart in energy as for protons. Thus, only two imperfection resonances, and no intrinsic resonances, exist in the AGS below 30 GeV.

Polarization in storage rings is feasible if the polarization in the steady-state does not decrease too fast. The simple theory above does not predict any depolarization; however, scattering, beam-beam interactions, field and energy fluctuations, etc. might conceivably produce slow depolarization. (Experiments at ANL, so far, have shown no evidence of such effects, but have been limited to short storage times.)

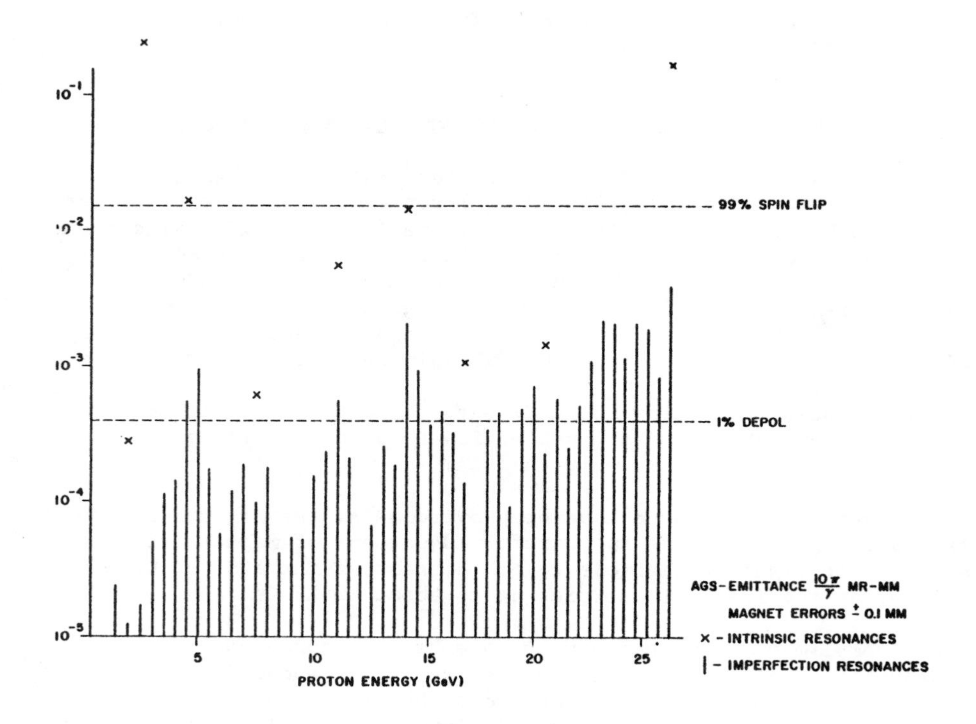

Fig. 1.

DEPOLARIZATION DURING ACCELERATION AND STORAGE OF POLARIZED PROTONS AND DEUTERONS

Chairman E.D. Courant; Co-Chairman L.G. Ratner

I. Acceleration of polarized protons to energies greater than 12 GeV/c.

   A. Problems associated with AG machines and boosters

      1. Resonances are stronger and there are more of then than at weak focusing ZGS.

         a. Conventional corrections as at ZGS of different depolarizing resonances

            i. Intrinsic
            ii. Imperfection
            iii. Other

         b. Cooling techniques and/or reduction of phase oscillations

         c. Removal of misalignments by improved surveying

         d. New ideas

      2. Relative merit of present machines

         a. Only the AGS has no booster

         b. Machines with boosters are PS, SPS, FNAL, (Serpukhov?)

         c. Depolarization due to magnet imperfections

         d. Intensities achievable considering sources and injection efficiency

         e. Set a reasonable upper limit for the polarized beam energy in each machines.

   B. Advantages compared to production from hyperons. (Intensity, polarization, energy and angle spread)

II. Acceleration of polarized deuterons

   A. Problems associated with deuteron acceleration.

      1. Suitability of LINACS and boosters for good intensity.

      2. Manipulation of rf for the heavier ion.

      3. Measurement of polarization of deuterons or stripped protons

      4. Efficiency of stripping and tagging

B. Advantages

1. Many fewer depolarizing resonances

2. A polarized neutron beam

III. Storage and colliding polarized beams

A. ISR-problems and advantages

1. Complicated system with booster, PS, and ISR but asymmetry parameters in the PS range are well known.

2. Estimate the possible luminosity

3. Depolarization during storage for many hours.
(Calculations, experiments)

4. Efficiency of stripping and tagging polarized deuterons for proton injection into the ISR.

B. ISABELLE

1. The AGS without booster is a simpler system.

2. Problems of acceleration in ISABELLE without depolarization
(Run ISABELLE at 30 x 30 GeV?)

C. Other semi-colliders like Fermilab schemes

IV. Accelerator research on depolarization using the ZGS Polarized Proton Beam

A. Deuteron acceleration and stripping

B. Cooling of polarized protons

C. Storage of polarized protons and/or deuterons.

D. Detailed behavior of intrinsic and imperfection resonances.

E. Others

V. Acceleration schemes with polarization in accelerator plane and rotating (studied by Derbenev, Kondratenko, and Skrinsky).

TOPICS TO BE STUDIED FOR STORING A POLARIZED BEAM

L.C. Teng
Fermilab
September 19, 1977

We assume that a polarized beam has been accelerated to the final energy for storage. Possible depolarizing processes during acceleration are studied elsewhere in the Workshop.

A. During injection and stacking

1. Extraction from accelerator, Transport to storage ring, Injection into storage ring - These are well understood problems, but should be taken into account in the design so that the polarization is preserved all the way through.

2. Betatron stacking (multi-turn injection) - This is also a well understood process and should not be a problem.

3. Stacking by charge exchange injection - Polarized negative ions can be injected by charge exchange injection. The effect of the stripping foil should be investigated, but the depolarizing effect is expected to be small or negligible.

4. Momentum stacking - In an AG storage ring the depolarizing resonances, especially the error resonances, are generally very closely packed. So even the small amount of acceleration required for momentum stacking may take the beam across several resonances. This should be studied in detail.

B. During storage

In general, any oscillatory electromagnetic field sensed by the ions may cause depolarization if it has a frequency component in resonance with the transition frequency $\Delta E/h$ between the up and down states (precession frequency). The source of the EM field includes:

ISSN: 0094-243X/78/103/$1.50

1. Guide magnetic field - This is the well known intrinsic and error resonances given by $\frac{g-2}{2}\gamma = n \pm m\nu$. In an AG storage ring the error resonances are rather densely packed. It may be difficult for the stacked beam with a large momentum spread to avoid straddling a resonance.

In addition, ripples in the guide field power supply will add to the richness of the frequency components and hence increase the number of resonances.

2. RF field - If the stored beam is to be kept bunched by the RF. The frequency components of the RF field should be investigated.

3. Beam self-field - The beam itself will induce EM fields in the vacuum chamber. Depending on the structure of the vacuum chamber the self-induced field could have rather complex frequency components.

4. Residual ion field - Ions produced by beam-ionization of the residual gas in the vacuum chamber may be trapped in the beam and produce a depolarizing field. This possibility, although remote, should be looked at.

C. Stochastic processes

1. Multiple Coulomb scattering by residual gas - The possibility of a spin-flip interaction is likely to be very small, but this should be confirmed.

2. Intra-beam scattering - The probability of a spin-flip interaction between 2 beam ions is also expected to be small. Again, this should be looked into.

THEORY OF SPIN DEPENDENCE AT VERY HIGH ENERGIES*

Gerald H. Thomas
Argonne National Laboratory, Argonne, Illinois 60439

ABSTRACT

The point of view I wish to take in this talk is to introduce a number of topics for possible discussion during the course of this workshop. In the first half of the talk, I will review some of the theoretical expectations for polarization phenomena assuming that hadrons are composite. In the second half of the talk I will consider polarization phenomena from an S-matrix point of view. The summary talk of Francis Low contains the conclusions which were reached during the course of this workshop.

HADRONS ARE COMPOSITE?

A popular notion is that hadrons are composite, and their constituents are spin 1/2 quarks and spin 1 gluons. The view is that the theory of strong interactions is an SU_3 gauge theory of color gluons, interacting with quarks which carry one of three color charges. This theory, Quantum Chromo-Dynamics (QCD)[1] is very similar in structure to Quantum Electro-Dynamics (QED). In such theories, the interaction of the fermions with the vector bosons is not arbitrary but fixed by the requirements of Lorentz invariance and local gauge invariance. If the theories are correct, then the observed spin dependence of the interaction must obey certain laws. In particular, in regions where single vector exchange dominates the interaction, the spin dependence of the interaction is expected to be quite simple. In what follows we will only consider processes where this is thought to be the case.

At the level of single vector gluon exchanges, QED and QCD are identical except for color factors. For scattering processes involving hadrons we must use some theory of composite object scattering to extract the underlying constituent interaction. (Possibly because of the SU_3 color symmetry, free quarks are not available for direct experimentation.) Experimenters are requested to think of the analogous problem of studying neutron interactions; one usually needs the Glauber theory to disentangle the underlying neutron-hadron interaction from the more accessible deuteron-hadron processes. For quarks, the theory which allows us to disentangle the quark interactions from hadron processes is the parton model. This model is reviewed in many places[2] so we won't say more about it at this time. We shall concentrate instead on the underlying processes, in particular the spin structure of the fermion-fermion scattering mediated by vector exchange and various instances of this in lepton-hadron and hadron-hadron scattering.

---

*Work performed under the auspices of the United States Department of Energy.

ISSN: 0094-243X/78/105/$1.50

A. Spinology

The discussion that follows will be at the level of the textbook of Bjorken and Drell[3]. The reader can refer to it for more details. The process we shall consider is sketched in Fig. 1.

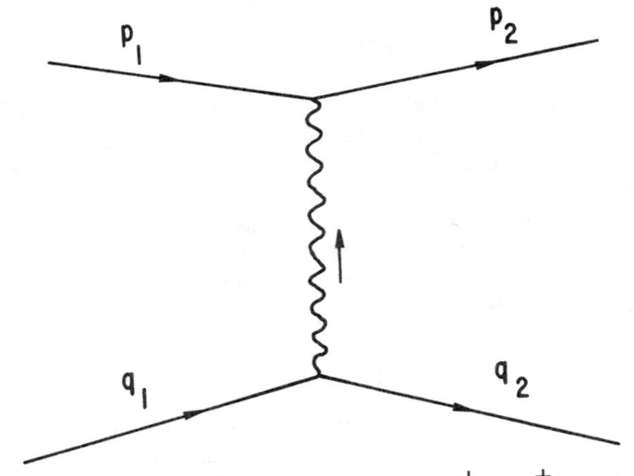

Fig. 1. One of the Feynman diagrams for $1/2^+$ $1/2^+$ elastic scattering mediated by 1^- exchange.

We shall be interested in the structure of this diagram, which follows directly from an application of the Feynman rules.

$$M = \bar{u}(p_2)\gamma_\mu u(p_1) g^{\mu\nu} \bar{u}(q_2)\gamma_\nu u(q_1) \tag{1}$$

is proportional to the invariant amplitude for Fig. 1. We have dropped factors of i and e^2 as well as the momentum dependence of the vector propagator since we do not intend to use (1) to compute absolute rates, but rather wish to highlight the spin dependence. This spin dependence comes in through the fermion wave functions

$$u(p) = \sqrt{E+m}\begin{pmatrix} \chi \\ \frac{\sigma \cdot p}{E+m}\chi \end{pmatrix} \tag{2}$$

where χ is a two component spinor. The fermion is characterized by both a momentum 4-vector p_μ and a polarization 4-vector w_μ: in the rest frame $p_\mu = (m,0,0,0)$ and $w_\mu = (0,\hat{w}_x, \hat{w}_y, \hat{w}_z)$ with $\hat{w}\cdot\sigma\chi = \chi$. For computations with Dirac wave functions, it is convenient to recall the covariant projection operators

$$u(p)\,\bar{u}(p) = \frac{(\not{p}+m)(1+\gamma_5\not{w})}{2} \; . \tag{3}$$

Using this fact we can evaluate the absolute square of M [Eq. (1)].

To start with we assume only p_1 and p_2 are polarized. For unpolarized particles, we sum (3) over both polarization states w_μ and $-w_\mu$

$$\sum_{\text{spins}} u(q)\bar{u}(q) = (\not{q} + m) \ . \tag{4}$$

The resultant probability is

$$\sum_{\text{spins}} |M|^2 = \sum_{\text{spins}} \bar{u}(p_2)\gamma_\mu u(p_1)\ \bar{u}(p_1)\gamma_\rho\ u(p_2)$$
$$\times \sum_{\text{spins}} \bar{u}(q_2)\gamma^\mu\ u(q_1)\ \bar{u}\ (q_1)\gamma^\rho\ u(q_2) \ . \tag{5}$$

Using the projection operators in Eq. 3-4, we find

$$\sum_{\text{spins}} |M|^2 = \mathrm{Tr}\left\{ (\not{p}_2 + m)\frac{(1+\gamma_5\not{w}_2)}{2}\gamma_\mu(\not{p}_1 + m)\frac{(1+\gamma_5\not{w}_1)}{2}\gamma_\rho \right\}$$
$$\times\ \mathrm{Tr}\left\{ (q_2 + m)\gamma^\mu(\not{q}_1 + m)\gamma^\rho \right\} \ . \tag{6}$$

This has the form of the product of two tensors

$$\sum |M|^2 = \mathrm{TOP}_{\mu\rho}\ \ \mathrm{BOTTOM}^{\mu\rho} \ , \tag{7}$$

one associated with the top vertex of Fig. 1, the other the bottom vertex. Of course this decomposition would remain true if particles q_1 and q_2 were also polarized; then the two tensors would have the same forms. The possible spin structure of the interaction follows from considering only the top vertex.

If the fermions are transversely polarized, and their masses are negligible, the form of the top vertex simplifies to

$$\mathrm{TOP}_{\mu\nu} = \frac{1}{4}\ \mathrm{Tr}\left\{ \not{p}_2(1+\gamma_5\not{w}_2)\gamma_\mu\not{p}_1(1+\gamma_5\not{w}_1)\gamma_\nu \right\} \tag{8}$$
$$= S_{\mu\nu} + D_{\mu\nu}$$

where

$$S_{\mu\nu} = \frac{1}{4}\ \mathrm{Tr}(\not{p}_2\gamma_\mu\not{p}_1\gamma_\nu)$$
$$= p_{1\mu}p_{2\nu} + p_{1\nu}p_{2\mu} - p_1 \cdot p_2 g_{\mu\nu} \tag{9}$$
$$= S_{\nu\mu}$$

and

$$D_{\mu\nu} = \frac{1}{4}\,\mathrm{Tr}(\not p_2 \not w_2 \gamma_\mu \not p_1 \not w_1 \gamma_\nu)$$

$$= D_{\nu\mu} \qquad . \tag{10}$$

There are no terms in $TOP_{\mu\nu}$ proprotional to a single fermion spin.

If the fermions are longitudinally polarized and the masses are negligible, we can replace $(1+\gamma_5 \not w)$ by the helicity projection operator $(1+\lambda\gamma_5)$ for a helicity state $\lambda = \pm 1$. The form of the top vertex again simplifies to

$$TOP_{\mu\nu} = \frac{1}{4}\,\mathrm{Tr}\left\{ \not p_2 (1+\lambda_2\gamma_5)\gamma_\mu \not p_1 (1+\lambda_1\gamma_5)\gamma_\nu \right\}$$

$$= (1+\lambda_1\lambda_2) S_{\mu\nu} + (\lambda_1+\lambda_2) A_{\mu\nu} \tag{11}$$

where $S_{\mu\nu}$ is as before [Eq.9] and

$$A_{\mu\nu} = -\frac{1}{4}\,\mathrm{Tr}\left\{ \gamma_5 \not p_2 \gamma_\mu \not p_1 \gamma_\nu \right\}$$

$$= i\,\varepsilon_{\alpha\beta\mu\nu} p_2^\alpha p_1^\beta = -A_{\nu\mu} \qquad . \tag{12}$$

Equation (11) shows the interaction conserves helicity, and furthermore allows a spin dependence if only one of the fermions is polarized. In this case, Parity requires the other vertex to have one fermion with definite helicity.

If both transverse and longitudinal spins are allowed, we can combine (8) and (11) to obtain

$$TOP_{\mu\nu} = (1+\lambda_1\lambda_2) S_{\mu\nu} + (\lambda_1+\lambda_2) A_{\mu\nu} + D_{\mu\nu} \ , \tag{13}$$

since one can easily show there are no interference terms between longitudinal and transverse polarization states.

We draw the following conclusions from [Eq. 13]:

(1). If only one fermion is polarized at one vertex, the interactions will be independent of its polarization unless one fermion at the other vertex is also polarized. When one fermion at each vertex is polarized, the interaction will be independent of their polarizations unless both fermions have longitudinal polarizations.

(2). When only longitudinal polarizations are considered, each vertex conserves the helicity of the fermion [i.e. $\lambda_1 = \lambda_2$ in (Eq.13) or the vertex vanishes].

(3). Transverse spin dependence occurs only when both fermions at one [or both] vertices have non zero transverse polarization.

In the discussions that follow, we consider only longitudinal polarizations, in which case we have helicity conservation at each vertex. We do not exclude however, interesting effects for transverse polarized fermions. There is the effect (4) above, as well as the possibility of interchange effects [e.g. the cross diagram to Fig. 1] causing non-zero dependence on the transverse spins.

As a final comment on spinology, for those of you who are familiar with the $pp \to pp$ Jacob and Wick helicity amplitudes[4,5]

$$\begin{aligned}
\phi_1 &= \langle ++ |\phi| ++ \rangle \\
\phi_2 &= \langle -- |\phi| ++ \rangle \\
\phi_3 &= \langle +- |\phi| +- \rangle \\
\phi_4 &= \langle +- |\phi| -+ \rangle \\
\phi_5 &= \langle ++ |\phi| +- \rangle \quad ,
\end{aligned} \tag{14}$$

I quote their value, for the Feynman diagram of Fig. 1, up to an overall numerical constant

$$\begin{aligned}
\phi_1 &= \frac{s}{t} \\
\phi_2 &= 0 \\
\phi_3 &= -\frac{u}{t} \\
\phi_4 &= 0 \\
\phi_5 &= 0 \quad .
\end{aligned} \tag{15a}$$

If the crossed graph is added, recalling that the fermions are identical, the full expressions are

$$\begin{aligned}
\phi_1 &= s\left(\frac{1}{t} + \frac{1}{u}\right) \\
\phi_2 &= 0 \\
\phi_3 &= -\frac{u}{t} \\
\phi_4 &= \frac{t}{u} \\
\phi_5 &= 0 \quad .
\end{aligned} \tag{15b}$$

One can then refer to the literature for the expressions of any spin observable in terms of these helicity amplitudes.[6]

B. Applications

With the formalities now out of the way we propose to give a qualitative description of several processes of possible interest, starting with deep inelastic electron scattering. One can refer to standard references[7] for the kinematics of the process (cf Fig. 2).

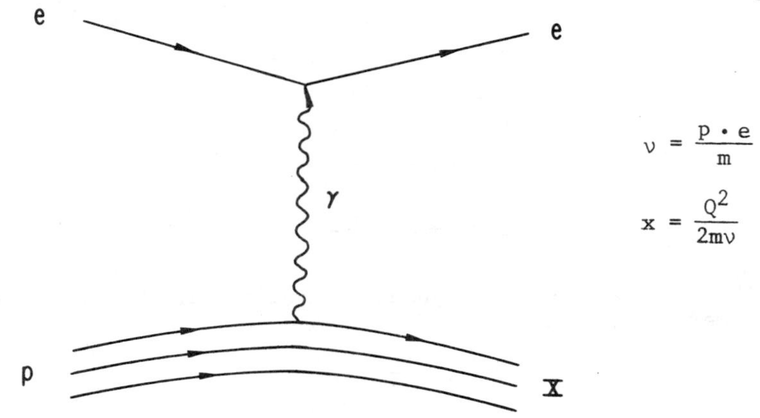

Fig. 2. Deep Inelastic Scattering: ep → eX where ν is the lab. energy of the incident electron, $q^2 = -Q^2$ is the 4-momentum transfer and x is the quark's fraction of the proton longitudinal momentum. The proton is viewed as three quarks (as is the final state X).

The elementary subprocess is electron quark elastic scattering via the exchange of a photon. If both incident fermions are in states of definite helicity, we would expect an asymmetry A_0 in comparing the cross section for aligned helicities against non-aligned:

$$A_0 = \frac{\sigma(++) - \sigma(+-)}{\sigma(++) + \sigma(+-)} \quad . \tag{16}$$

where $\sigma(\lambda_e \lambda_q)$ is the elementary (differential) cross section for an electron with helicity λ_e to scatter from a quark with helicity λ_q. Now we can prepare electrons in pure helicity states, but the quarks will in general not be 100% polarized since they are bound.

The quark parton argument[8] is that there is a certain probability to find a quark with the fraction x of the proton's longitudinal momentum, and with helicity λ_q given that the proton had definite helicity λ_p. This probability or structure function is a measure of the quark's wave function inside the proton. Because of this structure function, the quark will in general have a polarization P(x) along the direction of the proton's helicity, in addition to the usual probability G(x) for finding the quark.

Using these quite crude arguments, we expect an asymmetry A in $ep \to eX$ which is A_0 diluted by the quark polarization. More precisely, in the deep inelastic region

$$A = \frac{\sum_i Q_i^2 P_i(x) G_i(x)}{\sum_i Q_i^2 G_i(x)} A_0 \tag{17}$$

where the sum goes over the quarks in the proton and Q_i is the quark charge. If one is not at high q^2, there is an additional kinematic dilution in (17). For those wishing more details, I refer to an excellent review by R. Field.[9]

We now consider the Drell-Yan[10] process $p\bar{p} \to \mu^+\mu^- X$. The elementary process is the crossed reaction to Fig. 2: $q\bar{q} \to \mu^+\mu^-$ through a single photon. The kinematics are different, but the same structure functions enter. The expected asymmetry A for longitudinally polarized protons and antiprotons to annihilate into μ pairs plus anything is[8]

$$A = \frac{\sum_i Q_i^2 \int dx_1 dx_2 G_i(x_1)\bar{G}_i(x_2)\ \delta(M^2 - sx_1x_2) P_i(x_1) P_i(x_2) A_0}{\sum_i Q_i^2 \int dx_1 dx_2 G_i(x_1)\bar{G}_i(x_2)\ \delta(M^2 - sx_1x_2)} \tag{18}$$

where A_0 is the asymmetry for the elementary process. Thus A is the same as the elementary asymmetry modulo a dilution factor, and so A can in principle reflect the underlying dynamics.

Next we turn to the Drell-Yan processes $pp \to \mu^+\mu^- X$ (Fig. 3).

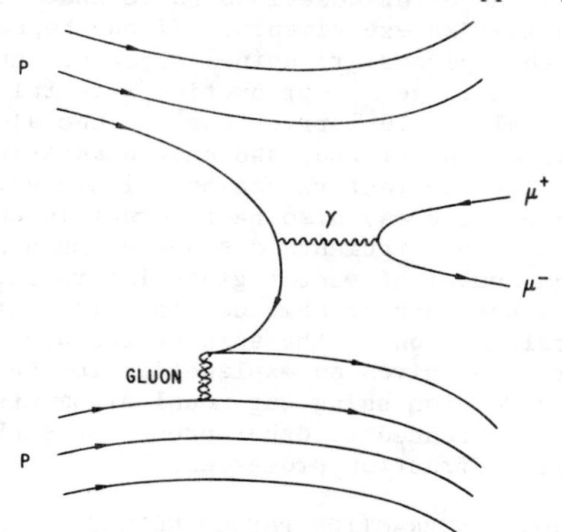

Fig. 3. Drell-Yan production of μ-pairs.

The only difference between this process and the last is that the antiquark must come from the sea. In order to see an effect, one must believe that a polarized proton causes the sea to be polarized.[11]

From graphs such as Fig. 3, this is possible. The asymmetry is otherwise the same as (18).[8]

The last example is large $P_\perp$ hadron production

$$\begin{aligned} &pp \to \pi X \\ &pp \to \text{jet } X \\ &pp \to \pi\pi X \\ &pp \to \text{jet jet } X \ . \end{aligned} \tag{19}$$

The only change from Fig. 3 is that the subprocess $q\bar{q} \to \mu^+\mu^-$ is replaced by $q\bar{q} \to q\bar{q}$ via a color vector gluon, and one includes in (18) the redressing probability for a quark to become a hadron (π, K, jet, etc.). Again the asymmetry measurement will reflect the elementary asymmetry modulo a dilution factor.

In summary we have seen qualitatively how the vector interaction might be studied through polarization measurements. The main effect of scattering composite objects is to dilute the effect. Now in presenting these arguments we have not been too careful with the form of the dilution factor. Indeed, the main difficulty lies in calculating such a factor since the pieces which make it up are not well known. One needs the u and d quark's polarization separately, as well as that of the sea quarks. At present, there is not enough experimental information from deep inelastic scattering data to determine these. Moreover, theoretical arguments are probably suspect though naive arguments do give about the right results for the asymmetry seen for A (Eq. 17) at SLAC.[12]

There are thus a number of questions to be answered before one considers these polarization experiments. If one hopes to establish in hadron reactions that gluons are spin-1 objects, one should first be in a region where the large $P_\perp$ production rate falls as $P_\perp^{-4}$ (modulo logarithmic, calculable corrections).[13] One also needs to establish that quarks are polarized, and know quantitatively how much so that the various dilution factors can be calculated.

The constituent picture may also be relevant in understanding soft collisions. The above arguments are not of much use since one certainly has a large number of vector gluon interactions instead of only one. Perhaps bags are of some use in this context. It is amusing that Low's calculation of the spin dependence of $pp \to pp$ near the forward direction gives an explanation for helicity conservation of the bare Pomeron using bag model arguments.[14] Perhaps these arguments can be extended to other processes such as $\Lambda p \to \Lambda p$, $\Lambda\Lambda \to \Lambda\Lambda$ or inelastic diffraction processes.

STRONG INTERACTION PHENOMENOLOGY

However much one likes the idea that the correct theory for all processes is a gauge field theory, one must face up to the difficulties of making quantitative predictions about purely strong interaction processes. We will here take the point of view that one way of learning about how to make such predictions is to study the data. There has developed a number of phenomenological models and ideas which have helped elucidate general properties of the S matrix, properties one might one day hope to establish from an underlying theory. Even if one never achieves such a connection with field theory, the study of the S-matrix may eventually lead to a satisfactory phenomenology of strong processes.

In this workshop we are focussing on spin, and what we might learn from studying how particles with spin interact. At energies below 100 GeV, the study of two body and quasi-two body processes has uncovered a vast number of regularities and near regularities of the S-matrix. There is every reason to believe that detailed studies at high energies will also reveal interesting physics.[15] So much has been written on S-matrix physics,[16] in this talk I need to do little more than mention a few highlights relevant for this workshop.

To start, let us recall some of the more obvious spin dependent interactions. In the previous section we looked at the consequences of a fermion-vector

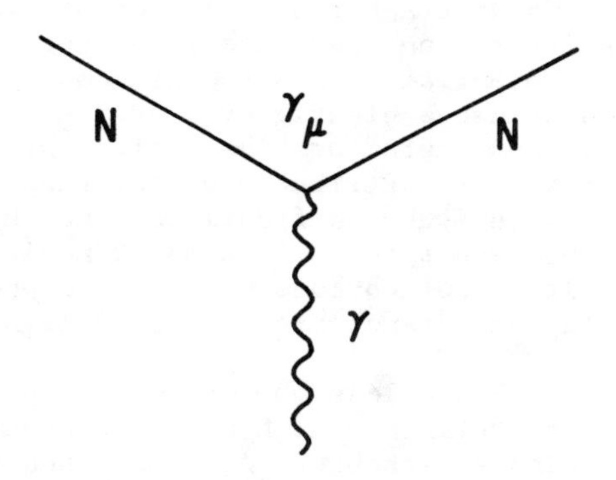

Fig. 4.

interaction. Another simple interaction is that of a nucleon with a pion at very low energies. To lowest order

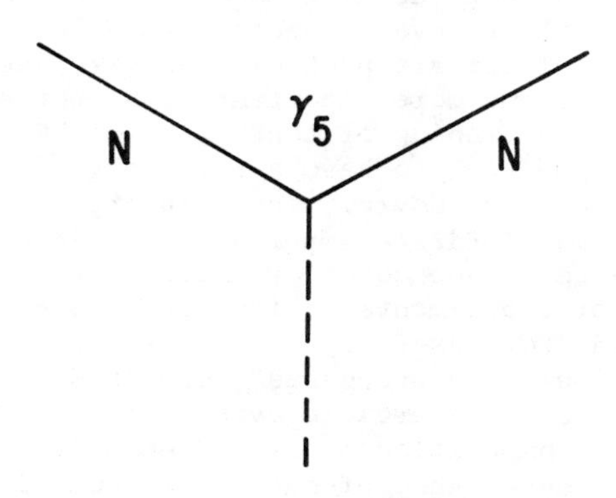

Fig. 5

the interaction is $\bar{u}\gamma_5 u$ and leads to a helicity flip interaction. One of the most distressing features of strong interactions is that such simple and obvious mechanisms are never exactly correct.

The simple couplings follow from requirements of Lorentz invariance and some notion of the fields interacting weakly. In the case of nucleons coupling to pions, the coupling $g^2/4\pi \approx 15$ is not weak. Nevertheless analysis of the high partial waves of low energy NN phase shifts shows evidence of elementary pion exchange. The more correct interpretation is in terms of the analyticity of the S-matrix: there is evidence for a t-channel singularity at the position of the pion, whose strength can be determined from the phase shift data, and is found to have a value $g^2/4\pi \approx 15$.

What is quite amazing is that there is also evidence in high energy processes for pion exchange. Reactions such as $np \to pn$[17] and $pp \to n\Delta^{++}$ [18], show evidence of the pion exchange pole, and all give $NN\pi$ couplings compatible with the low energy phase shift value.

Also significant is that both at low energies and at high energies the simple Feynman graph result is not precisely valid. For the NN phase shift region there are large correction terms to the one pion exchange potential at short distances. Only at large distances is the pion exchange clearly evident. At higher energies, one invokes the ubiquitous "cuts" or "absorption corrections" to modify all but the peripheral attributes of the pion exchange.

Why one needs to make such modifications, and why such a simple spin interaction can be seen after such drastic modifications is an interesting puzzle. It is not obvious what will happen to the pion exchange in the TeV region; it may prove to be enlightening and should be studied.

One has learned from data less obvious spin dependences.[19] For example, NN elastic scattering at high energies behaves as though the Pomeron is the dominant exchange. As determined from pp experiments, the helicity conserving amplitude is the one that dominates. There is a wealth of such regularities, and many reviews can be consulted for details.[20] What may concern us at this workshop is what are interesting things to look for as we go to higher energies.

One possibility is that the Regge and multiparticle phenomenology learned over the last several years extends naturally to the higher energy domain. There are problems however. Because Regge cuts are expected to become more important at higher energies, we should perhaps see the beginning of that trend at FNAL. For example, one should expect $\pi^- p \to \pi^\circ n$ to have a noticably different t-dependence at 200 GeV/c than at 20 GeV/c. The data are on the contrary rather simple. This may indicate important gaps in our understanding about cuts, and hence about even low energy physics.[21] Thus, bread and butter experiments on low t, two body and quasi-two body reactions should prove useful.

What about multiparticle processes? Our present phenomenology finds few departures in multiparticle events from the expectation that the process is a composition of many "few body, low subenergy" processes.[22] The spin dependence of such reactions might lead to progress in our understanding.[15]

We know the spin averaged total cross section rises through the ISR energy range. We might learn something about the underlying mechanism by studying whether pure spin cross sections such as $\Delta\sigma_T$ and $\Delta\sigma_L$ also rise at the higher energies.[23]

At very high energies, there have been speculations that strong processes might interfere with weak processes.[24] Spin is a good tool to use to find parity violation effects. One side benefit is that one might learn something about the absolute phase of the strong S-matrix in an analogous manner as one learns about phases from Coulomb interference. In the case of weak-strong interference, this might extend to very high $p_\perp$.

REFERENCES

1. A recent and somewhat theoretical review of QCD has been given by W. Marciano and H. Pagels, "Quantum Chromodynamics, a Review," Rockefeller University report number COO-2232B-130 (1977).
2. The physical basis for the parton model can be found in R. P. Feynman, Photon-Hadron Interactions, (W. A. Benjamin Inc., Reading Mass., 1972).
3. J. D. Bjorken and S. D. Drell, Relativistic Quantum Mechanics (McGraw-Hill Inc., New York, 1964), vol. 1.
4. M. Jacob and G. C. Wick, Ann. Phys. 7, 404 (1959).
5. M. L. Goldberger, M. T. Grisaru, S. N. MacDowell and D. Y. Wong, Phys. Rev. 120, 2250 (1960).
6. F. Halzen and G. H. Thomas, Phys. Rev. D10, 344 (1974).
7. A survey of deep inelastic scattering can be found in the proceedings of the Tbilisi Conference. The plenary review is by V. I. Zakharov, "Plenary Report on Deep Inelastic Scattering," Proceedings of the XVIII International Conference on High Energy Physics, p. B69 vol. II.
8. Our discussion follows J. Kuti and V. F. Weisskopf, Phys. Rev. D4, 3418 (1971).
9. R. D. Field, "Some Remarks about Large $P_\perp$ Spin Effects," Proceedings of the symposium on experiments using enriched antiproton, polarized-proton and polarized antiproton beams at Fermilab energies, June 19, 1977, ANL-HEP-CP-77-45, p. 88.
10. S. D. Drell and T-M Yan, Phys. Rev. Letters 25, 316 (1970); Annals of Physics 66, 578 (1971).
11. F. E. Close and D. Sivers, Phys. Rev. Letters 39, 1116 (1977).
12. M. J. Alguard et al., Phys. Rev. Letters 37, 1261 (1976).
13. Recent work which tries to accomodate the observed p_T^{-8} behavior of large p_T data with the expected scaling behavior of p_T^{-4} puts in scaling violations as calculated from asymptotic freedom, and puts in the transverse motion of partons. See G. Fox, review at the Coral Gables Conference 1978. Other work includes R. Cutler and D. Sivers, Phys. Rev. D16, 679 (1977).
14. F. E. Low, Phys. Rev. D12, 163 (1975).
15. C. Sorensen, "Some Remarks about Physics with Polarized Proton and Antiproton Beams (mostly soft processes)," Argonne Symposium June 10, 1977, ANL-HEP-CP-77-45, p. 210.

16. See for example the recent book by P. D. B. Collins An Introduction to Regge Theory and High Energy Physics, (Cambridge U. P. New York 1977).
17. See Particle Data Group, "NN and ND Interactions (above 0.5 GeV/c) - A Compilation," UCRL-20000 NN August 1970, p. 146 ff.
18. See Particle Data Group, op. cit., p. 86 ff. For polarization data, and some phenomenological discussions, see the review by R. Diebold, "Proceedings of the Conference High Energy Physics with Polarized Beams and Targets;" ed. M. L. Marshak (AlP, New York 1976), p. 92.
19. For recent analysis of low energy data see P. Kroll, E. Leader and W. von Schlippe, "Structure of the $I = 1$ nucleon-nucleon amplitudes at 6 GeV/c," Westfield college preprint WU-B-77-9 August 1977; E. L. Berger, A. C. Irving and C. Sorensen, "Implications of Nucleon-Nucleon Spin Polarization Measurements," ANL-HEP-PR-77-86 Nov. 1977.
20. E.g. C. Quigg and G. C. Fox, Ann. Rev. Nucl. Science 23, 219 (1973); G. L. Kane and A. Seidl, Rev.Mod. Phys. 48, 309 (1976).
21. G. Farmelo and A. C. Irving, Nucl. Phys. B128, 343 (1977).
22. See for example G. H. Thomas, "Inclusive Correlations in the Central Region," ANL-HEP-PR-77-01 Jan. 1977, and references therein contained.
23. See G. H. Thomas, "Inelasticity and Structure in $pp \to pp$ at Medium Energies," ANL-HEP-CP-77-57, August 1977.
24. E.g. E. Fischbach, "Hadronic Weak Interactions and the Parton Model at High $p_\perp$," Argonne polarization symposium June 10, 1977, ANL-HEP-CP-77-45, p. 65.

THEORY OF HIGH ENERGY SPIN DEPENDENCE

F.E. Low, Chairman
G.H. Thomas, Cochairman

I. Constituent Picture

A. Are partons polarized?
1. Drell-Yan processes
2. High p_T elastic scattering
3. High p_T inclusive hadron production

B. What is the fundamental force law between constituents?
1. QCD at short distances?
2. MIT Bags at longer distances?
3. Diffractive Excitation - Is there still helicity conservation?
4. What happens in $\Lambda\Lambda \rightarrow \Lambda\Lambda$? What is the spin orbit coupling?

II. S-Matrix Picture

A. What is the S-matrix?
1. Elastic diffraction scattering.
2. Inelastic diffraction scattering
3. Two body or quasi-two body scattering
4. Central region inclusive production
5. Fragmentation region inclusive production
6. Multiparticle production

B. Areas where spin dependence might settle specific issues
1. Rising total cross section
2. Short range correlations of inclusive production
3. Nature of diffraction
4. Duality for baryons
5. Parity violating processes in hadronic reactions

ISSN: 0094-243X/78/117/$1.50

EXPERIMENTS ON SPIN DEPENDENCE AT VERY HIGH ENERGY

M. L. Marshak
School of Physics, University of Minnesota
Minneapolis, Minnesota 55455

We review recent measurements of spin-dependent parameters in strong interactions. The focus of the review is those measurements which suggest the possible behavior of these parameters in future experiments at very high energy.

INTRODUCTION

The purpose of this conference is to answer the questions "What experiments can we do on spin effects at very high energy?" and "What experiments should we do?" In neither case is the answer obvious. In this paper, we shall attempt to point in the right direction by reviewing what has been done to study spin effects in high energy interactions in recent years. Because of the scope of the conference, we shall limit this review to strong interactions, although the question of spin effects in the weak and electromagnetic interactions is also a very interesting subject [1].

IS SPIN IMPORTANT AT HIGH ENERGIES?

Although the very existence of this conference is a partial answer to that question, we should convince ourselves before proceeding that spin effects are indeed important at very high energies (> 100 GeV). In fact, our intuitions generally point in the opposite direction. With an increasing amount of linear momentum present in the interaction, how is it possible that a constant amount of angular momentum can have a large effect. It is necessary when answering this question to attempt to distinguish between "important" and "large". Certainly, large effects are important. Small effects can be important as well, but given our limited understanding of this field, it is probably best to concentrate on large effects first.

Elastic pp scattering is the most familiar reaction in which to measure spin effects. The experimentally simplest spin-dependent measurements in that reaction are the polarization by the scattering of an unpolarized beam (polarizing power) and the analyzing power (left-right scattering asymmetry) of a polarized beam or target, all of which are constrained to be equal. These data are shown in Fig. 1 as a function of s (the c.m. energy

ISSN: 0094-243X/78/118/$1.50

squared) for scattering at small t (the four-momentum transfer squared) [2]. The figure shows that this spin-dependent parameter clearly tends towards zero with

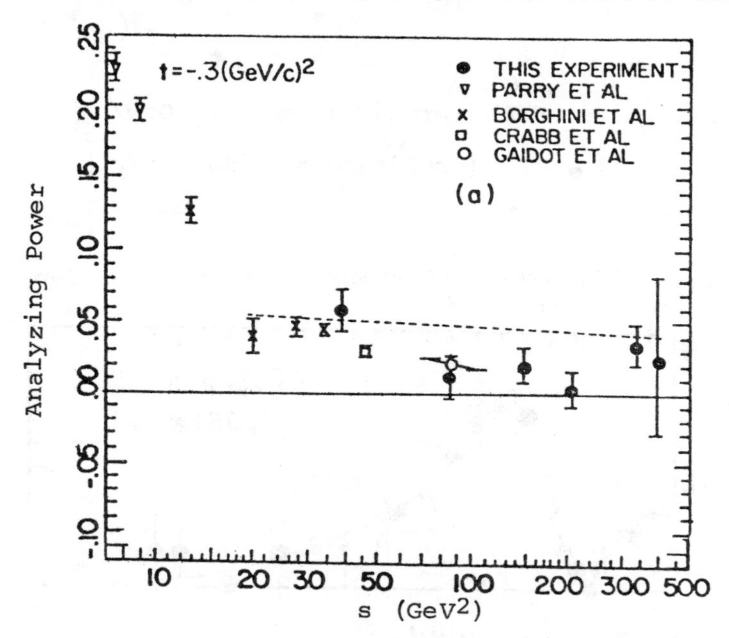

increasing energy. The results of the most recent measurements at Fermilab and the CERN SPS [3,4] are shown in Fig. 2 with analyzing power at fixed s plotted as a function of t. Although there is some spread between the different experiments (the CERN data are very preliminary), the analyzing power at small t generally appears small at high energies.

This small analyzing power does not necessarily mean that all spin effects in high energy pp elastic scattering are small. In this reaction there are four spins; measurements have been made of some of the spin-correlation parameters of the reaction. The highest energy measurement of a spin-correlation parameter in pp elastic scattering is that of A(NN) at 11.75 GeV/c [5]. These data, shown in Fig. 3, emphasize two important effects. A(NN) at 12 GeV/c is large (30 percent), particularly at large t. A(NN) is very different (and in general larger) at 12 GeV/c than at 6 GeV/c. What will result when A(NN) is measured at 100 GeV/c is pure conjecture, but in Fig. 1, 12 GeV/c ($s \approx 25$) is very close to the asymptotic behavior region.

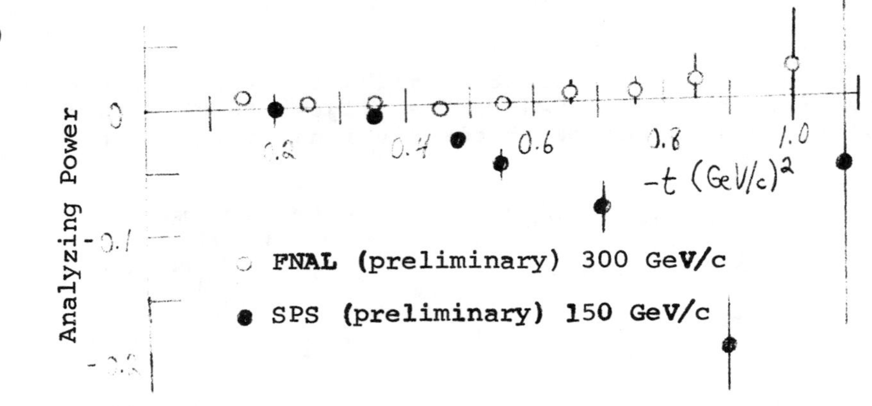

Fig. 2--The analyzing power in high energy pp scattering for $|t| \lesssim 1.0$ $(GeV/c)^2$ from Refs. 3 and 4.

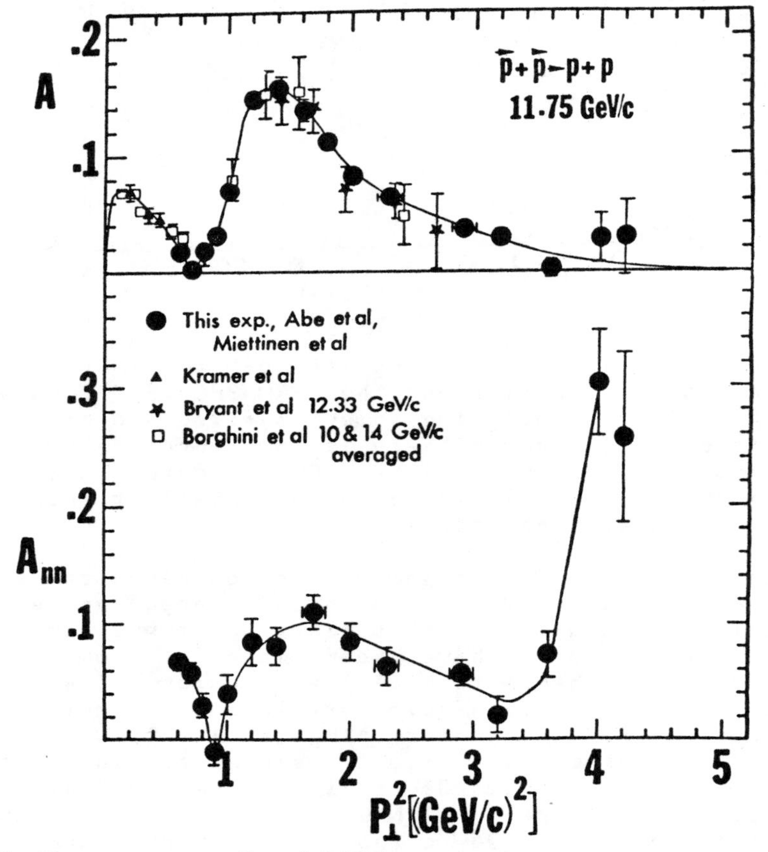

Fig. 3--The parameters A and A(NN) in pp elastic scattering at 11.75 GeV/c from Ref. 5.

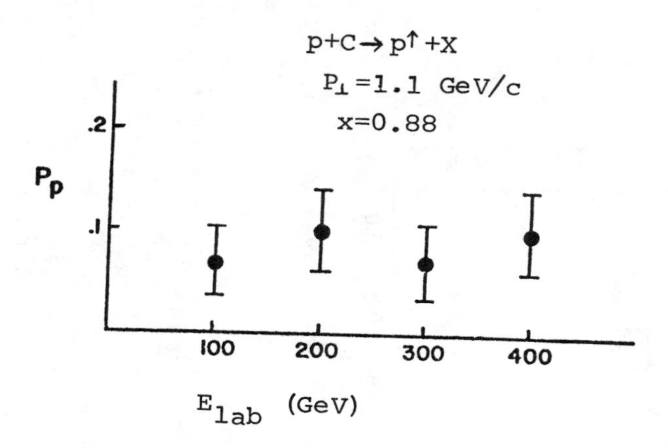

Fig. 4--The polarization of the scattered proton in inclusive p-carbon scattering as a function of lab energy. (Ref. 6)

Recent results from "jet" experiments suggest that high energy inclusive processes may be a better probe of the basic quark-quark interaction than any other type of scattering. If this conjecture is true, we should be able to study the spin dependence of the qq interaction by measuring the spin dependence of inclusive processes such as $p + p \rightarrow p + X$, where only one particle is detected in the final state. A similar measurement has been made at the Fermilab internal target region [6], with the results shown in Fig. 4. Here we find that a single spin parameter in the inclusive reaction $p + C \rightarrow p\uparrow + X$ has a moderately large (10 percent) but more importantly an energy independent value between 100 and 400 GeV.

What really happens at the next energy frontier is always somewhat unknown, but there is no clear evidence that spin is unimportant at high energy and some evidence that spin may strongly affect the underlying quark-quark interaction. The only way to really find out is to do the experiments. In the remainder of this review, we shall consider the data from previous experiments as a guide to determining what spin-dependent experiments we should do at high energy.

DISCUSSION OF RECENT DATA

In the case of pp elastic scattering, there are five independent amplitudes, which means that nine independent parameters (five magnitudes and four phases) can be determined. (The overall phase can only be measured at small t by interference with the Coulomb amplitude.) When the problem is stated in this manner, our knowledge of pp

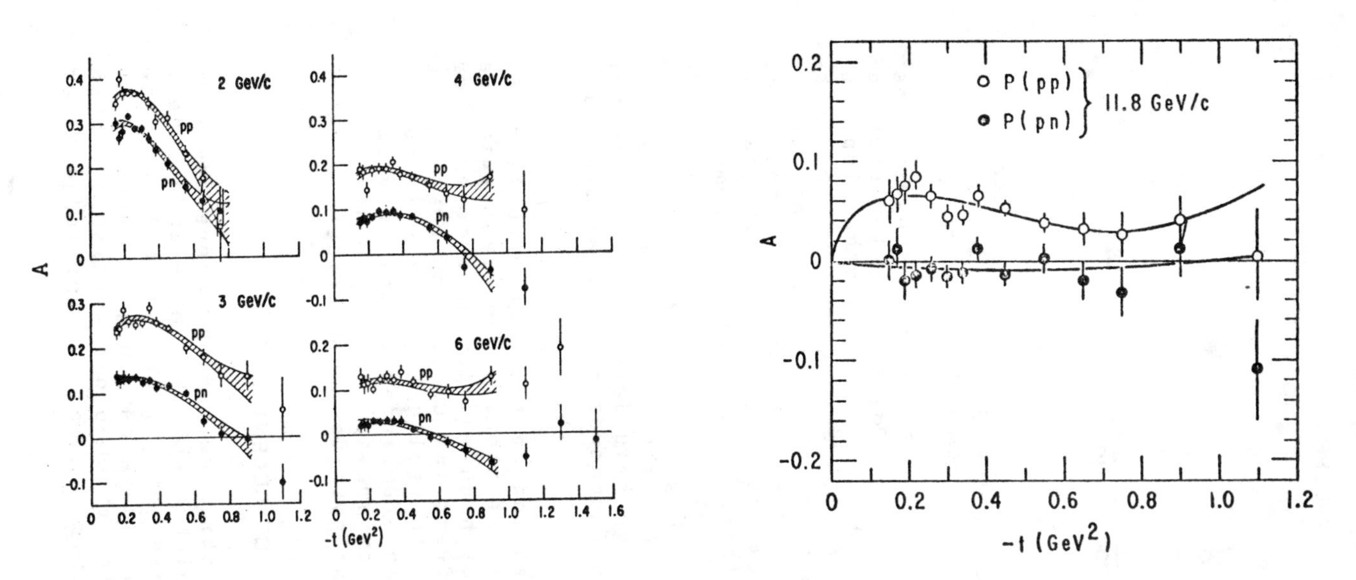

Fig. 5--The analyzing power (often called polarization) in pp and pn elastic scattering from 2 to 6 GeV/c (from Ref. 8)

Fig. 6--(above right) The analyzing power for pp and pn elastic scattering at 11.8 GeV/c and (below right) the pn elastic analyzing power at 3 GeV/c at large |t| (Ref. 9).

elastic scattering appears quite limited. Complete phase shift solutions do not currently exist for scattering at more than 1 GeV of kinetic energy. Data collected within the last few years at the Argonne ZGS, however, should permit a complete amplitude analysis for moderate t at 6 GeV/c [7]. Above this energy, our knowledge is currently limited to cross-sections, polarizations (or equivalent analyzing powers) and A(NN) (or equivalent cross-section difference) measurements at 11.75 GeV/c. Our knowledge of spin effects in the isospin-related pn elastic process is even worse. Figs. 5 and 6 show recent data at $|t| < 1.5$ (GeV/c)**2 for the analyzing power in pn elastic scattering [8], compared to pp data in the same kinematic range. The pn analyzing power at small t appears to vanish with increasing incident energy even faster than the pp analyzing power. Data at 24 GeV/c which are not shown are substantially similar to the 12 GeV/c data. There is a hint of negative analyzing powers at larger values of t, which is supported by backward scattering analyzing power measurements. However, there are currently no high energy 90° c.m. analyzing power data for pn scattering, except for some preliminary measurements [9] shown in Fig. 6. All of these data still leave unresolved the question of whether there is a constant difference between the pp and pn analyzing powers, independent of energy, as suggested by the data in the figures. From isospin invariance, one would expect that at high energies the analyzing powers for pp and pn scattering would be equal at least at small t. Any other result would require a detailed explanation for the difference.

We now turn to the spin-dependent data for other elastic processes. Again, at high energies, only target analyzing power data exist for $|t| < 1.2$ (GeV/c)**2. The π^+p and π^-p data [10] in Fig. 7 show the mirror symmetry, which has been seen before at lower energies.

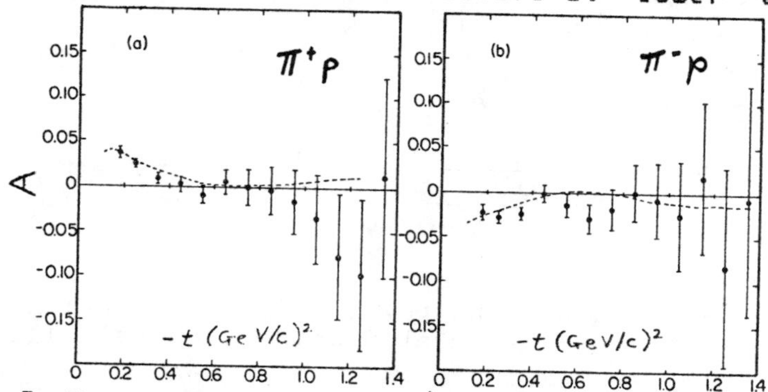

Fig. 7--The analyzing power in $\pi^{\pm}p$ elastic scattering at an incident momentum of 100 GeV/c from Ref. 10.

As in the pp case, the analyzing powers for both pion charges seem to approach zero with increasing energy. In the case of πN scattering, there are no data to indicate the behavior of this parameter at large t.

Much work in high energy physics has been devoted to the study of exclusive inelastic channels. However, the question of spin dependence in particle production has hardly been touched. There have been some measurements of production asymmetries [11] for the Δ^{++}, at a range of incident energies as shown in Fig. 8. Recent measurements at the Argonne ZGS have also determined the production asymmetry in the reactions $p\uparrow + p \rightarrow d + \pi$ and $d + \rho$ [12]. These data are interesting because the asymmetries for these two mesons are so different. The plot in Fig. 9 suggests that spin may be useful to separate particles from background in missing mass searches. It is also possible that the interference between distinct particles and the background can yield valuable information about those particles. There has also been one measurement at 6 GeV/c of the D parameter in pp inelastic scattering [13]. The difference between

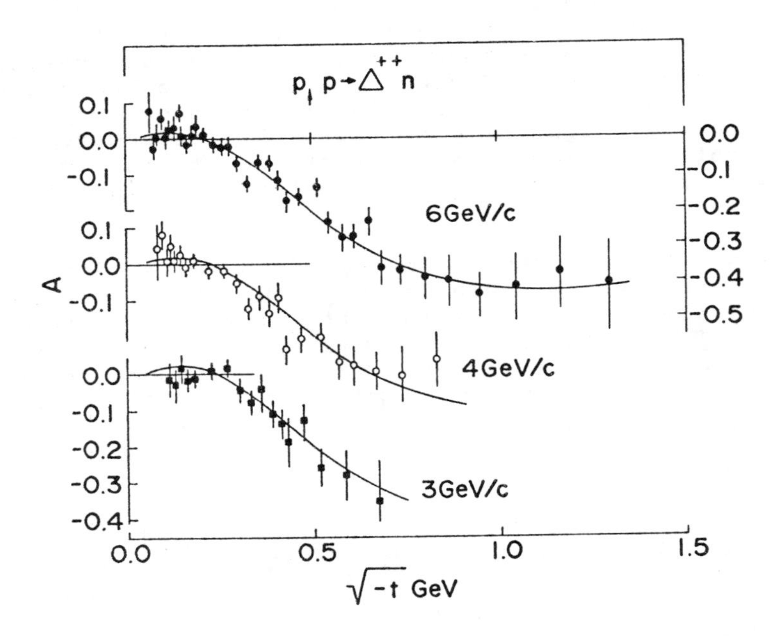

Fig. 8- The analyzing power A in the reaction $p + p \rightarrow \Delta^{++} + n$ at 6 GeV/c from ref. 11.

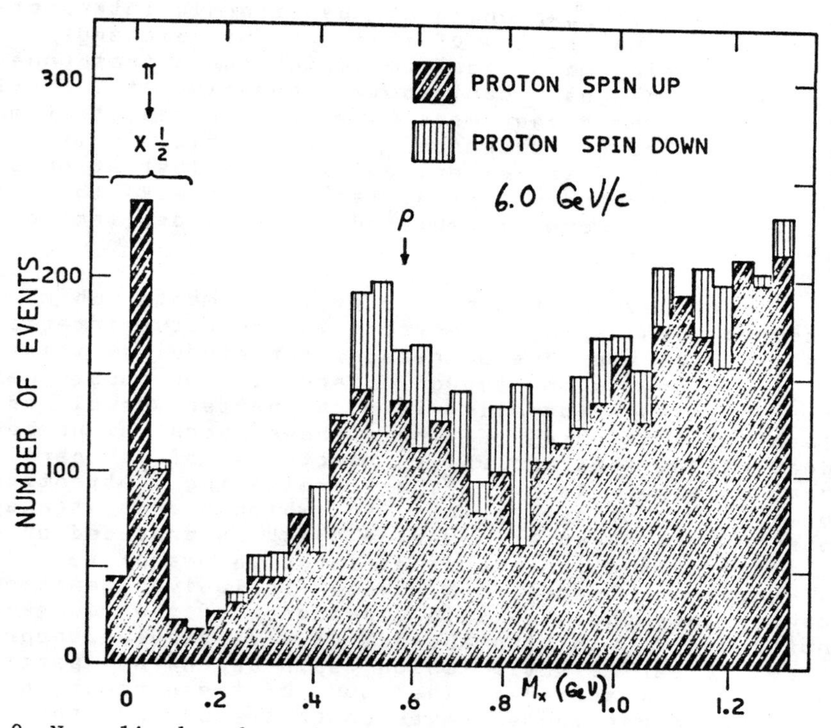

Fig. 9--Normalized number of events for the reaction $p\uparrow + p \rightarrow d + X$ as a function of M_x from Ref. 12.

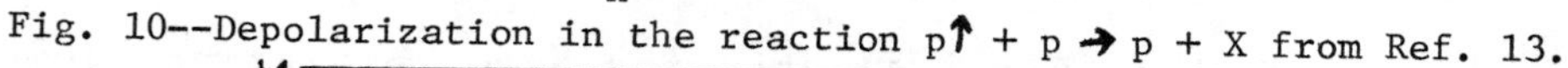

Fig. 10--Depolarization in the reaction $p\uparrow + p \rightarrow p + X$ from Ref. 13.

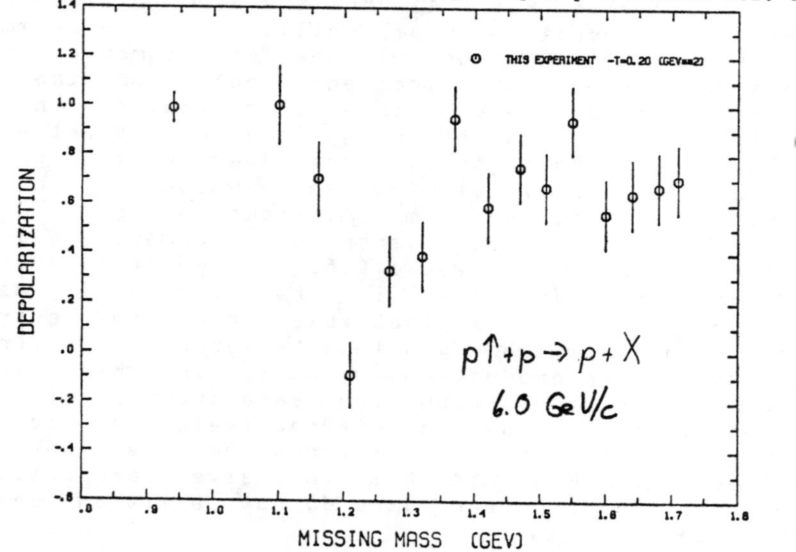

this parameter and unity can be directly interpreted in terms of the naturality of the t channel exchange. Fig. 10 is a missing mass spectrum, which shows that unnatural parity exchanges contribute heavily to N*(1232) production, the higher mass N*'s are the result of mostly natural parity exchange (presumably diffractive production). This experiment is an illustration of the use of spin-dependent parameters to check theoretical models which were formulated from cross-section data alone.

There have been a number of experiments which have measured spin dependent effects in inclusive interactions [14]. There are several reasons for studying this type of reaction, even though there is no doubt that the spin-dependence formalism is much better developed for elastic scattering. There have been a number of speculations that inclusive reactions of a particular type correspond to quark-quark scattering. Attention has focussed on those events in which large transverse momenta are observed, where a "jet" is produced or where particle production occurs at large values of x_F. What we hope to learn from these inclusive reactions is nothing less than the spin dependence of the quark-quark interaction. In the absence of a particular theory, it is not clear exactly which spin-dependent parameters should be measured in the study of these reactions. So far the experiments have concentrated on the beam analyzing power and the polarizing power, where it can be determined by the decay of an unstable particle.

The data in Fig. 11 show the beam analyzing power for pion production in $p^{\uparrow}p$ collisions. The asymmetry is plotted as a function of the four-momentum transfer squared between the incident proton and the outgoing pion. There are several interesting features in the data including an apparent energy independence between 6 and 12 GeV/c and the suggestion that features at high x_F also appear at low x at the same values of u, but in considerably diluted form. As shown in Fig. 12, these data have some resemblance to backward π p elastic scattering data at 6 GeV/c [15]. A general description of the inclusive data is that interesting asymmetry features at high x (possibly due to quark-quark scattering) are submerged in a background of diffractive, isotropic pion production at low x. In this view, the inclusive proton production data shown in Fig. 13 are uninteresting because the diffractively produced protons completely obscure the interesting reactions at all values of x. Fig. 14 shows inclusive K production data but the large error bars do not permit any definitive statements.

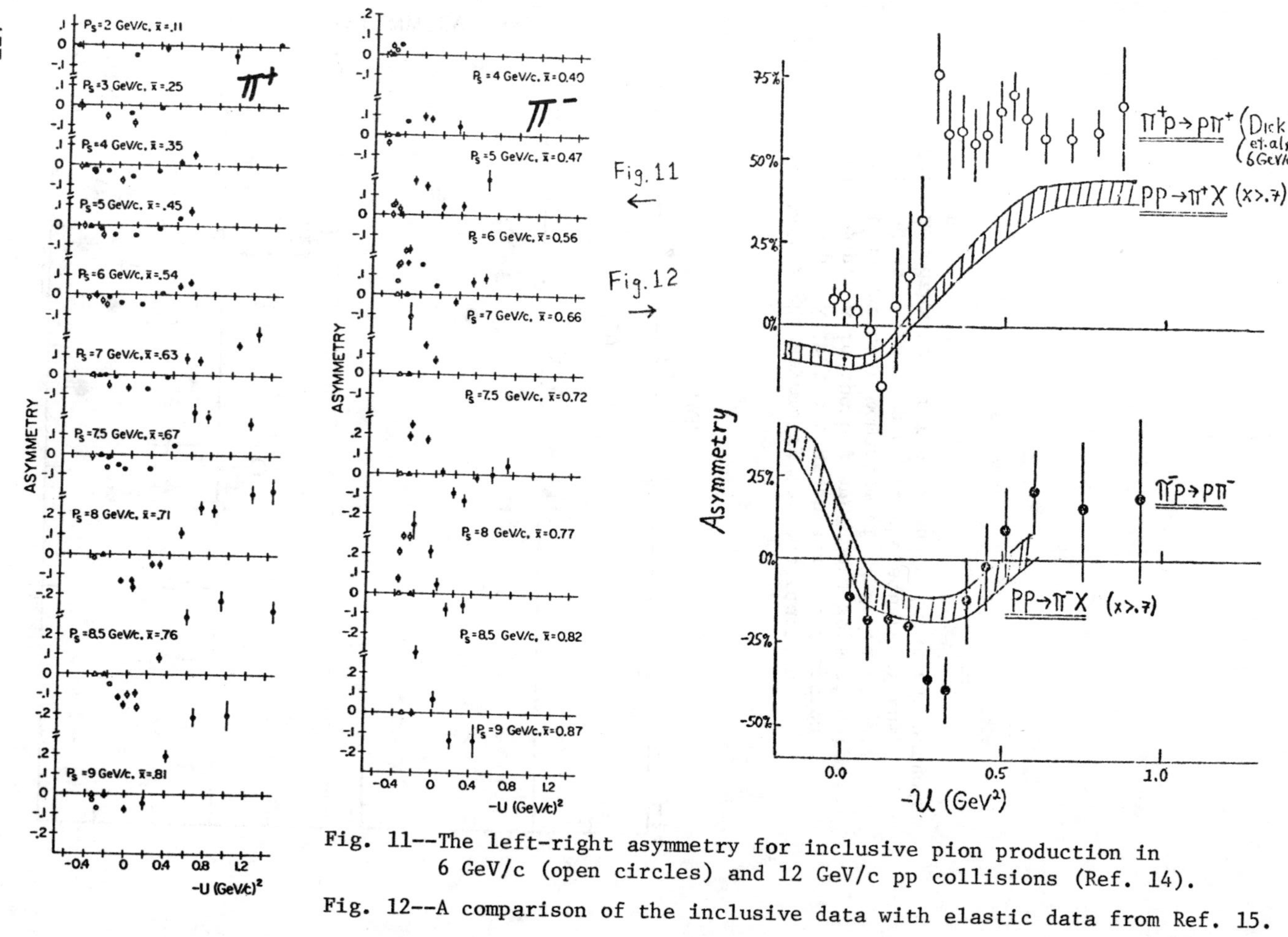

Fig. 11--The left-right asymmetry for inclusive pion production in 6 GeV/c (open circles) and 12 GeV/c pp collisions (Ref. 14).

Fig. 12--A comparison of the inclusive data with elastic data from Ref. 15.

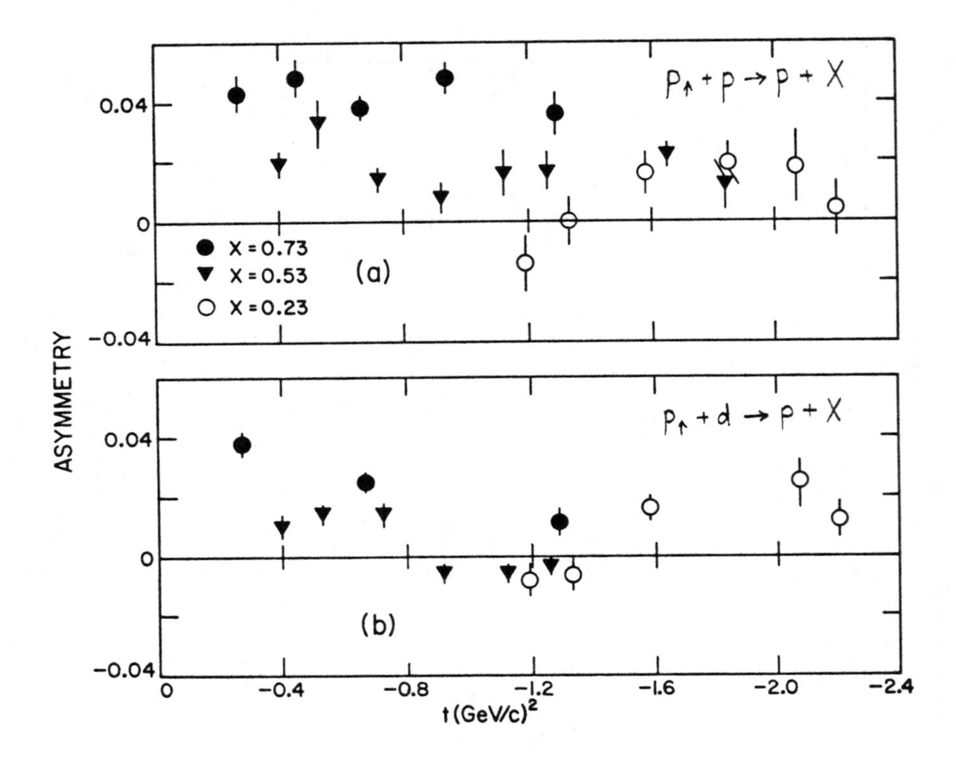

Fig. 13--The left-right asymmetry for the reaction $p\uparrow + p \rightarrow p + X$ in the a graph and the reaction $p\uparrow + d \rightarrow p + X$ in the b graph. Both sets of data are at an incident momentum of 12 GeV/c and are plotted as a function of t, the square of the four-momentum transfer between incident and outgoing protons.

Some very high energy studies of inclusive reactions have been made in the case of lambdas [16]. These experiments have the advantage that the Λ^0 is self-analyzing through its decay. Fig. 15 shows the final state polarization as a function of the transverse momentum. The selection of large $P_\perp$ seems to emphasize the interesting aspects of the phenomenum. Here again it is possible that the small $P_\perp$ and the large $P_\perp$ processes are quite different, with the former being isotropic and the latter emphasizing quark-quark scattering events.

SOME THOUGHTS FOR THE FUTURE

The absence of large amounts of data about high energy spin effects leaves room for much speculation. Committing these thoughts to paper is unfortunate, because the history of high energy physics shows that they will certainly be wrong. However, contemporary ideas will control the next round of experiments. For that reason alone, they have some value. Further details of these ideas are discussed in the summary report of this working group.

In elastic scattering, we must examine analyzing powers at high energy at large t. There is strong reason to believe that low t elastic scattering is primarily diffractive. For that reason and because some data already exist, low t is probably uninteresting. At large t, an extrapolation of existing data suggests that even the analyzing power at 100 GeV/c could show complex structure. There is certainly no evidence to suggest that we are anywhere near an asymptotic energy for $A(NN)$ at large t. There are also a number of other two-spin measurements that could be very interesting at large t. With a high energy polarized beam, measurements of the depolarization and spin rotation parameters at large t could yield a considerable amount of additional information about the mechanisms for these rare elastic scatterings.

The primary goal for the study of inclusive reactions is to test whether they are indeed, in certain kinematic regions, a manifestation of quark-quark scattering. Answering this question will require one-spin asymmetry measurements over a wider kinematic range and two-spin measurements (where both initial state protons are polarized) at high $P_\perp$ and high x. These data should be compared with asymmetries for "jet" production with polarized beams and targets. From these results some consistent picture could emerge about the nature of the spin dependence of the strong interaction.

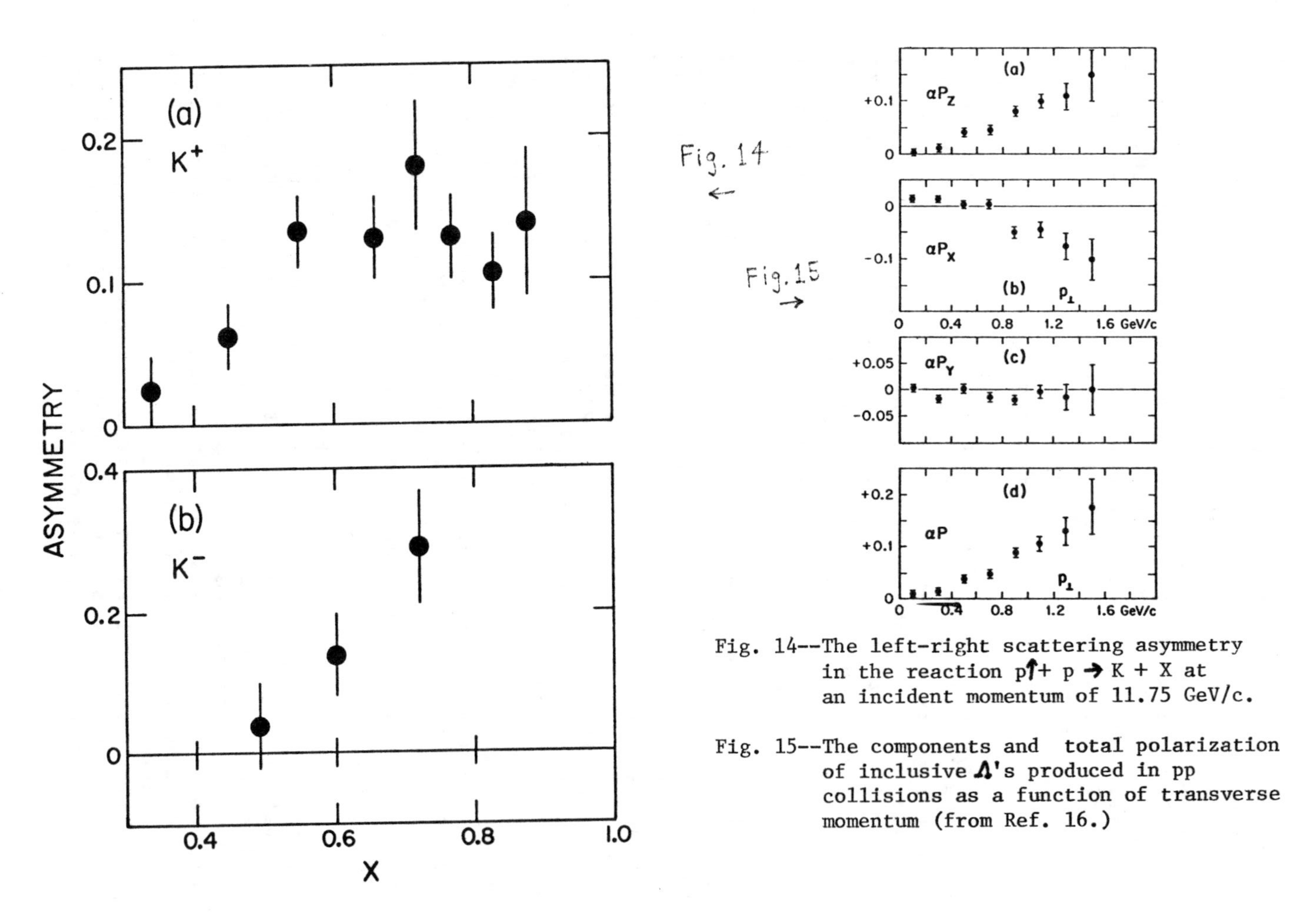

Fig. 14--The left-right scattering asymmetry in the reaction $p\uparrow + p \rightarrow K + X$ at an incident momentum of 11.75 GeV/c.

Fig. 15--The components and total polarization of inclusive Λ's produced in pp collisions as a function of transverse momentum (from Ref. 16.)

This work has been supported by the U. S. Department of Energy.

REFERENCES

1. For a review of recent work in the spin dependence of electromagnetic and weak interactions, see High Energy Physics with Polarized Beams and Targets, M. L. Marshak, ed., Amer. Inst. of Phys., New York, 1976.
2. M. Corcoran et al., Indiana University preprint IUHEE 16.
3. The FERMILAB data were presented at this conference by O. Chamberlain.
4. The SPS data were presented at this conference by G. Fidecaro.
5. J. R. O'Fallon et al., Phys. Rev. Lett. 37, 733 (1977).
6. M. Corcoran et al., Indiana University preprint.
7. These experiments are being conducted at the Argonne ZGS by a group headed by A. Yokosawa.
8. R. Diebold et al., Phys. Rev. Lett. 35, 632 (1975)
9. Preliminary data from the Argonne-Minnesota-Rice collaboration have been provided by E. Peterson.
10. I. P. Auer et al., Phys. Rev. Lett. 39, 313 (1977).
11. B. Wicklund in High Energy Physics with Polarized Beams and Targets, M. L. Marshak, ed., Amer. Inst. of Phys., New York, 1976, p. 198.
12. R. Klem et al., sumbitted to Phys. Rev. D.
13. H. Kagan, Ph.D. dissertation, University of Minnesota, 1978.
14. R. Klem et al., Phys. Rev. Lett. 36, 929 (1976).
15. M. Borghini et al., Phys. Lett. 31B, 405 (1970).
16. G. Bunce et al., Phys. Rev. Lett. 36, 1113 (1976).

EXPERIMENTS ON HIGH ENERGY SPIN DEPENDENCE

by M.L. Marshak, Co-chairman
[Chairman O. Chamberlain]

1. Is spin important at high energies?

   A. Elastic polarizations (analyzing power)
   Tend toward zero like 1/p at small t

   B. Spin correlations between incident and outgoing particles in elastic scattering. Large effects seen at the highest measured energies.

   C. Inelastic processes
   Polarization for p+C inelastic scattering appear energy independent between 100 and 400 GeV/c

   Conclusion: Spin is still an important quantum number at high energies.

2. Discussion of recent data

   A. Elastic scattering

      1. pp--polarizations, spin correlation measurements

      2. pn--how does the difference between pp and pn change with energy?

      3. Other elastic processes--πN scattering, p-nucleus scattering

   B. Exclusive inelastic processes
   $pp \rightarrow p\Delta$, $d\pi$, $d\rho$
   Is spin relevant to particle spectroscopy?

   C. Inclusive reactions

      1. What do we learn from inclusive processes?

      2. What quantities should we measure?

      3. π, K and p production from nucleons

      4. Λ production

      5. Production from nuclear targets

3. Some thoughts for the future

   A. Elastic analyzing power at large t

   B. Elastic correlation parameters

   C. Detailed study of why inclusive spin effects are large

   D. New particle searches

   (Speculation: Can cooled $\bar{p}$ storage rings have polarized beams?)

ISSN: 0094-243X/78/132/$1.50

POLARIZED PROTON BEAMS PRODUCED BY HYPERON DECAY AT VERY HIGH ENERGY

Chairman G. Fidecaro

The possibility of setting up a polarized proton beam from Λ° decay was considered for the first time by Overseth at the 1969 NAL Summer Study. Overseth came to the conclusion that a High Energy proton beam with intensity 10^5 protons/pulse and 50% polarization appeared feasible.

In 1972 Dalpiaz et al., reconsidered this possibility for the CERN SPS EPB and concluded that one could expect to get up to 10^6 protons/pulse.

In 1976 Atherton and Doble considered the feasibility of a polarized proton beam at a standard hadron beam line (H_2 beam) in the SPS north area by making use of the kinematical restriction in Λ° decay. The intensities expected are now between $5 \cdot 10^6$ and 10^6 polarized proton/pulse, with a polarization varying between 30 and 60%, in the momentum range 300-350 GeV/c. At the experiment target all protons are confined within a circle of less than 20 mm diameter. The horizontal angular divergence is $x' = \pm 0.6$ mrad, and the vertical divergence is $y' = \pm 0.4$ mrad. The momentum spread $\Delta P/P$ is less than 2%. The main background source ($K^\circ \rightarrow \pi^+\pi^-$ giving a pion in the beam) results in a pion contamination of less than 10^{-4}. The same beam could provide also polarized antiprotons but at 300 GeV the intensity would be several orders of magnitude smaller.

Some groups have examined the possibility of using the polarized proton beam in combination with polarized and unpolarized proton targets, for an extensive program of measurements (SPS proposal P87).

ISSN: 0094-243X/78/133/$1.50

Recently polarized beams from hyperon decay are also being considered at Fermilab.

Although at present the only practical way to get high energy polarized protons seems to be the one outlined above, one should be aware that the intensity (or better, the square root of intensity times the polarization) appears to be the main limiting factor.

POLARIMETER STUDY GROUP

J.B. Roberts - Chairman

I. Proton Polarimeters

A. Low Energy

1. After ion source (~20 KV)
Atomic/very low energy nuclear methods

2. After pre-accelerator (~1 MeV)
Exothermic nuclear reactions - solid state detectors

3. After Linac (50-500 MeV)
p-nucleus elastic scattering: single arm high resolution detectors, double arm right-left spectrometers

B. High Energy

1. After Booster (1-8 GeV)
Absolute polarimeters (p-p elastic scattering): simple double arm spectrometers, high resolution single arm
Relative polarimeters: p-C, p-CH_2 recoil polarimeters

2. After Main Ring (20-400 GeV) (fixed target)
Small t pp elastic scattering: double arm spectrometer, high resolution single arm, jet target in ring
Inclusive scattering polarimeters: pion production, inelastic proton scattering, inclusive lambda polarimeter

3. Storage Ring Polarimeters
Small t elastic scattering: pulsed atomic beam target

II. Deuteron Polarimeters
Same outline as proton polarimeters: at low energies (<1 GeV), use p-d elastic scattering. At high energies, use p-d elastic scattering. At high energies, use $p+(pn)\rightarrow pp$ with double arm spectrometers from liquid hydrogen target or CH_2, or inclusive scattering from fixed targets.

ISSN: 0094-243X/78/135/$1.50

POLARIZED ION SOURCES AND LOW ENERGY STORAGE

Hilton F. Glavish, chairman
ANAC, inc

I. POLARIZED ION SOURCES

The two types of polarized ion source widely known and extensively used over the past decade are:

1. The Atomic Beam Source, and

2. The Lamb-Shift Source

Motivated by the greed high energy physicists have for higher and higher beam currents, remarkable developments in the performance of atomic beam sources have occurred during the past two to three years. One of the topics to be discussed in this workshop is the nature of these developments, and perhaps even more important, the potential future developments that can now be reasonably projected.

While there have been significant developments in recent times, ideas for atomic beam sources have been substantially exhausted. Therefore it is important that a portion of the effort at this workshop should be devoted to imaginative thinking (in the first approximation), and imaginative thinking consistent with realistic budgets (in the second approximation). As long as there exists an ultimate goal to conduct experiments with monochromatic beams in the sense of spin state as well as momentum state, there is a need to develop intense polarized beams. This will be the case even if clever techniques are used for injection and acceleration, and even if low energy storage rings with or without electron cooling, are utilized in some way.

To some of us, the recent developments in atomic beam sources are fairly well known. Fortunately, this workshop is timely enough that we do not need to be so monotonous as to just re-iterate the technical details, the limitations, and the potentials of atomic beam sources. For now there is a third type of source:

3. The Colliding Beam Polarized Ion Source,

which is far less well known, but promises a real breakthrough in beam current and beam intensity.

A summary of the status and performance of each source is given below.

ISSN: 0094-243X/78/136/$1.50

1. ATOMIC BEAM SOURCE

Argonne National Laboratory first discovered[1] that the beam current from their source, available for injection into the ZGS, could be enhanced by a factor of 3 simply by pulsing the dissociator. The explanation for this surprising gain was not entirely clear, but was thought to be related to the cooler dissociator temperature which prevails in pulsed operation, and the lower residual gas pressures along the beam path just downstream of the nozzle. No matter what the explanation, it is an experimental fact that pulsing increased the beam current from 8 microamps H^+ dc to 25 microamps H^+ pulsed.

Following the improvements made at the ZGS in both the polarized source and the handling of the beam during injection and acceleration, plus the definitive demonstration that depolarizing resonances could be successfully jumped, the feasibility of accelerating a polarized beam in a strong focusing synchrotron soon received attention.

Thus CERN began a program in conjunction with ANAC to build a more intense polarized source so that the beam currents during acceleration in the CERN PS would be sufficiently high to enable routine diagnostics to be conducted, including polarization measurements after resonance jumps. The work on the CERN prototype polarized source was completed early in 1977. The beam currents obtained[3] for H^+ and D^+ were:

CERN polarized beam current   50 microamps (dc)

The substantial improvement in beam current is attributed to:

1. Smaller drift space between the nozzle and sextupole,

2. Improved pumping in the region of the nozzle and sextupole entrance,

3. Higher sextupole magnetic fields and the addition of a compressor sextupole,

4. A longer ionizer with the provision to shape the magnetic field to produce a higher, stable electron density.

It is estimated that the last change produced a factor of 3 improvement and the compounded effect of the first three changes a factor of 2 improvement.

It may be noted that the enhancement from pulsing, in the CERN source, is less than in the ZGS source. This is probably because of the improved pumping in the CERN source. It should also be mentioned that in the meantime, some of the above changes have also been implemented on the ZGS source, producing 65 microamps pulsed (dc currents not available[2]).

The group at Bonn have found that very strong magnetic field in the ionizer achieved with a superconducting solenoid also enhances the beam current.

As a result of the work carried out with the CERN source, the effects of dissociator pulsing are now better understood. The increased beam current comes not from cooling, but rather from the instantaneous pressure increment generated by the initial dissociation of hydrogen molecules: $H_2 = H + H$.

2. LAMB-SHIFT SOURCE

For the last 6 or 7 years the Lamb-Shift source has produced the highest negative polarized beam currents. The small emittence that can be obtained with these cources has made them attractive for use with tandem Van de Graaff accelerators in nuclear physics. However, the total beam current of only 0.6 microamps, has not made this source attractive for synchrotron injection and, furthermore, little progress has been made in the last few years in enhancing the beam current.

For negative beams, the Colliding Beam Source, discussed in the next section shows much more promise.

3. THE COLLIDING BEAM POLARIZED ION SOURCE

In 1968 it was suggested by Haeberli[4] that large polarized beam currents might be obtained when a fast beam and a thermal polarized hydrogen beam collide.

At the University of Wisconsin the first prototype source of this type has been built and tested. Polarized H^- or D^- ions are generated from a colliding Cs° beam of 40 keV energy. The schematic arrangement is shown in Fig. 1. A porous, heated, tungsten button produces Cs^+ ions by surface ionization. These are accelerated to 40 keV and passed through a cesium neutralizer canal to produce fast Cs° atoms.

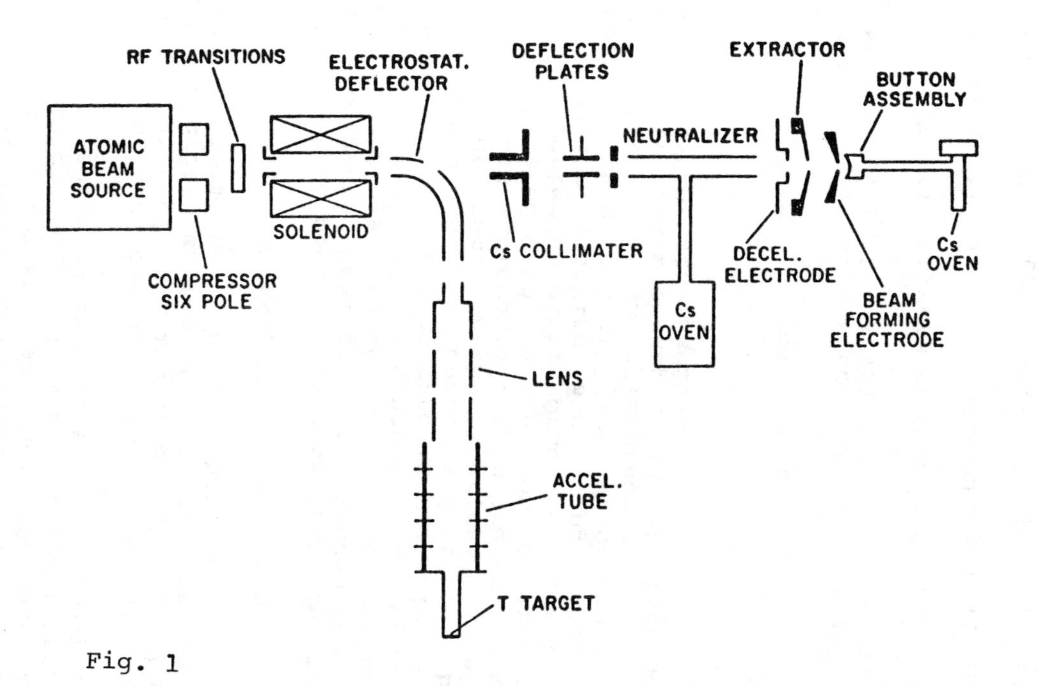

Fig. 1

The components of the University of Wisconsin colliding beam source. The source produces polarized H^- and D^- ions by colliding a 40 keV Cs^0 beam with the polarized neutral atomic beam. A Hot porous tungsten button produces Cs^+ ions by surface ionization of Cs metal. The Cs^+ ions are accelerated to 40 keV and then neutralized in a Cs vapor canal. The neutralized beam passes into a solenoid. Interaction with the atomic beam produces negative ions which are extracted and electrostatically deflected through 90^0.

Beam currents of 1.5 microamps of H^- or D^- ions have been extracted and mass analyzed. With improvements in the cesium gun design, one can reasonably project beam currents as high as 10 microamps.

Since H^- ions can be stripped for injection, a 10 microamp source of negative ions might well be equivalent to a 500 microamp source of positive ions, as far as accelerated beam currents are concerned.

II. LOW ENERGY STORAGE RINGS

It might be possible to substantially increase the final, accelerated beam current, through the use of a low energy storage ring which is filled during the acceleration cycle of the synchrotron. Depolarization in the low energy storage ring is probably small since resonances would be well spaced and it would be possible to sit between them. On the other hand, if electron cooling is also implemented, spin precessing magnetic fields associated with the electron confinement would have to be contended with. Other features to consider are: the vacuum, stacking, and timing. Also, there is the possibility of using H^- ions which are stripped on injection into a low energy storage ring. It would be worthwhile if this workshop could establish a reasonably definitive set of alternatives.

REFERENCES

1. E.F. Parker, N.Q. Sesol, and R.E. Timm, IEEE Trans. Nuc. Sci., NS-22 (1975) 1718.

2. E.F. Parker, private communication.

3. W. Kubischta and H.F. Glavish, private communication.

4. W. Haeberli, Nucl. Instr. and Methods, 62, (1968) 355.

APPENDICES

CONVENTION FOR SPIN PARAMETERS IN HIGH ENERGY SCATTERING EXPERIMENTS

Modified Version of October 24, 1977
Report of Notation Committee

[Ashkin, Leader-Chairman, Marshak, Roberts, Soffer, Thomas]

Approved by workshop on Oct. 25, 1977 with one dissention

I. Elastic and Pseudo-elastic Scattering

Consider the process

$$a+b \rightarrow c+d$$

where all four particles a,b,c, and d have spin 1/2 or less. We will denote the four particles in the following ordered way:

(a,b; c,d)
(beam, target; scattered, recoil)
(i,j; k,ℓ)

where each index, such as i, may take values such as ↑ (spin up) or ↓ (spin down), [or → (spin along momentum)]. We will conform to the Basel convention in that the forward-scattered particle, c, goes to the left, and the normal to the scattering plane is defined by:

$$\vec{N} = \frac{\vec{p}_a \times \vec{p}_c}{|\vec{p}_a \times \vec{p}_c|}$$

a) Fundamental Observables for Fixed Target Experiments in the Laboratory

Each spin i,j,k,ℓ can lie along the $\vec{N}$,$\vec{L}$, or $\vec{S}$ directions which are defined in LAB as follows:

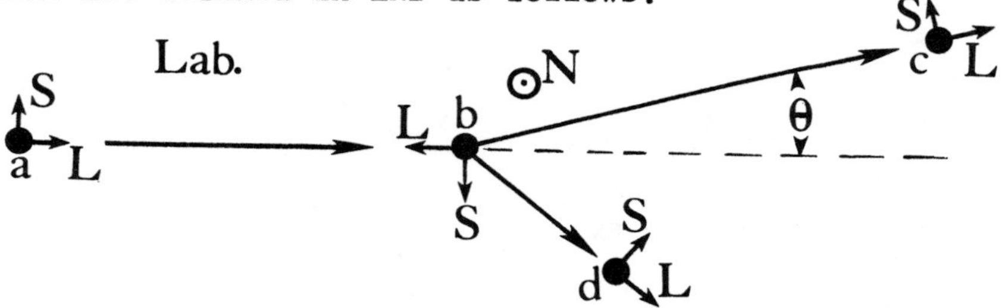

ISSN: 0094-243X/78/142/$1.50

Notice that $\vec{N}$ is out of the paper for all particles, and that for each particle, $\vec{L}$ points along the particle's momentum.† Note that $\vec{S}$ is defined by $\vec{S} = \vec{N} \times \vec{L}$.

b) Fundamental Observables in the CM or for Colliding Beam Experiments

Each spin $(i,j,k,\ell)_{CM}$ can lie along the $\vec{n}, \vec{\ell}$, or $\vec{s}$ directions which are defined in CM as follows:

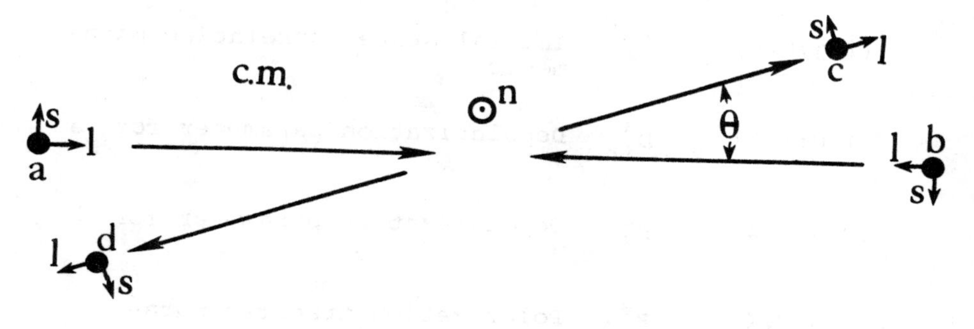

The direction of ℓ always points along each particle's momentum. Notice that while $\vec{N}$ and $\vec{n}$ are identical, $\vec{L} \neq \vec{\ell}$ and $\vec{S} \neq \vec{s}$. Therefore while spin-parameters like A_{nn} are invariant in going from lab to CM ($A_{nn} = A_{NN}$), parameters like $K_{\ell\ell}$ and D_{ss} are not invariant under this boost.

c) Polarization

"Polarization" refers to the state of a single particle or an ensemble and is short for "Degree of Polarization":

$P_B \equiv$ beam polarization

$P_T \equiv$ target polarization

$P_S \equiv$ scattered particle polarization

$P_R \equiv$ recoil particle polarization

d) Asymmetries Associated with One Polarized Particle*

We will use the 4-index notation to define these Asymmetries:

(n,0;0,0)	A^a	Analyzing power for a
(0,n;0,0)	A^b	Analyzing power for b

(0,0;n,0)	P^c	Polarizing power for c
(0,0;0,n)	P^d	Polarizing power for d

NOTE: 0 denotes unpolarized in initial state or polarization unmeasured in final state.

e) Correlations Associated with Two Particles Polarized*

(0,0;k,ℓ)	$C_{k\ell}$	Final state correlation parameter
(i,j;0,0)	A_{ij}	Initial state correlation parameter
(i,0;k,0)	D^a_{ik}	Depolarization parameter for a
(0,j;0,ℓ)	$D^b_{j\ell}$	Depolarization parameter for b
(i,0;0,ℓ)	$K^a_{i\ell}$	Polarization transfer parameter for a
(0,j;k,0)	K^b_{jk}	Polarization transfer parameter for b

NOTE: Each index i,j,k,ℓ refers to the spin orientation along either the $\vec{n}, \vec{\ell}$, or $\vec{s}$ direction. Thus A_{nn}, which was previously sometimes called C_{nn} is the initial state correlation when the spins are oriented along the normal to the scattering plane.

NOTE: $D^a_{ij} \neq D^b_{ij}$ for the $\vec{\ell}$ or $\vec{s}$ directions even for pp→pp

f) Recommend NO SPECIAL SYMBOLS for three or four polarized particle reactions. Just use 4-index notation such as (i,j;k,0) or (i,j;k,ℓ) or else use pure spin cross sections defined below.

g) If an experiment measures a mixture of observables care must be taken to state exactly what has been measured.

h) Differential Cross-section Measurements in Pure Initial and/or Final Spin States.

Label spin direction by arrows as indicated for the different pure initial spin cross sections

$$\text{n-polarization } \left.\frac{d\sigma}{dt}\right)_{\uparrow\uparrow}$$

$$\ell\text{-polarization } \left.\frac{d\sigma}{dt}\right)_{\rightleftarrows} \text{ or } \left.\frac{d\sigma}{dt}\right)_{\rightarrow\leftarrow}$$

$$\text{s-polarization } \left.\frac{d\sigma}{dt}\right)_{\otimes\odot}$$

These pure 2-spin (initial) cross sections are related to pure 3-spin and 4-spin cross sections by relations such as

$$\left.\frac{d\sigma}{dt}\right)_{\uparrow\uparrow} \equiv \left.\frac{d\sigma}{dt}\right)_{\uparrow\uparrow\rightarrow 00}$$

$$= \left.\frac{d\sigma}{dt}\right)_{\uparrow\uparrow\rightarrow 0\uparrow} + \left.\frac{d\sigma}{dt}\right)_{\uparrow\uparrow\rightarrow 0\downarrow}$$

$$= \left.\frac{d\sigma}{dt}\right)_{\uparrow\uparrow\rightarrow\uparrow\uparrow} + \left.\frac{d\sigma}{dt}\right)_{\uparrow\uparrow\rightarrow\uparrow\downarrow} + \left.\frac{d\sigma}{dt}\right)_{\uparrow\uparrow\rightarrow\downarrow\uparrow} + \left.\frac{d\sigma}{dt}\right)_{\uparrow\uparrow\rightarrow\downarrow\downarrow}$$

II. One Particle Inclusive Reaction

Consider the inclusive reaction

$$a+b \rightarrow c+X$$

where all three particles a,b, and c have spin 1/2 or less. The final particle c has CM momentum defined by

$$p_\perp \text{ and } x_c \equiv p_\parallel / p_c^{\,max}$$

All spin-parameters are defined as for the above elastic reactions. However care must be taken in defining the direction of the normal to the scattering plane

$$\hat{N} = \frac{\vec{p}_a \times \vec{p}_c}{|\vec{p}_a \times \vec{p}_c|}$$

that a,b, and c are defined so that particle c goes to the left. Notice that the spin-parameters may depend on the sign of x_c.

III. Total Cross-Section Measurements.

Arrows should indicate the spin directions of particles a and b in the process $a+b \rightarrow$ anything

$$\Delta\sigma_T \equiv \sigma_{\uparrow\downarrow} - \sigma_{\uparrow\uparrow} = \sigma_{\otimes\odot} - \sigma_{\otimes\otimes}$$

$$\Delta\sigma_L \equiv \sigma_{\substack{\rightarrow\\ \leftarrow}} - \sigma_{\substack{\rightarrow\\ \rightarrow}}$$

NOTE: First or top arrow refers to particle a

*The asymmetries (I-d) and correlations (I-e) are defined in terms of pure spin cross sections by relations such as:

$$P^c \equiv (0,0;n,0) \equiv \frac{\sigma_{00\rightarrow\uparrow 0} - \sigma_{00\rightarrow\downarrow 0}}{\sigma_{00\rightarrow\uparrow 0} + \sigma_{00\rightarrow\downarrow 0}}$$

$$A_{nn} \equiv (n,n;0,0) \equiv \frac{\sigma_{\uparrow\uparrow\rightarrow 00} + \sigma_{\downarrow\downarrow\rightarrow 00} - \sigma_{\uparrow\downarrow\rightarrow 00} - \sigma_{\downarrow\uparrow\rightarrow 00}}{\sigma_{\uparrow\uparrow\rightarrow 00} + \sigma_{\downarrow\downarrow\rightarrow 00} + \sigma_{\uparrow\downarrow\rightarrow 00} + \sigma_{\downarrow\uparrow\rightarrow 00}}$$

†In some experiments $\vec{L}$ for the target particle, b, was chosen to point along the beam direction. This convention causes considerable difficulty, as pointed out in Professor Michel's lecture, and we have decided to adopt the more logical convention, in spite of some strong historical feelings.

WHAT IS POLARIZATION? HOW TO COMPARE ITS MEASUREMENT WITH BEAM AND TARGET AND WITH COLLIDING BEAMS?

Louis Michel
IHES, 91440 Bures-sur-Yvette
France

Polarization is what has to be measured in order to completely specify the state of a particle whose energy and momentum are known. In quantum mechanics it has to be the expectation value of an operator commuting with E and $\vec{P}$, the energy and momentum operators. What is this operator? It was found probably by Pauli at least forty years ago and was known by oral tradition before it appeared in print (Lubanski, Physica 2 (1942) 310, quotes Pauli unpublished). Let us look for it.

Euclidian invariance of physics imposes momentum and angular momentum conservation. The commutation relations between the corresponding operators $\vec{P}$ and $\vec{J}$ are

$$[P_i, P_j] = 0, \quad [J_i, J_j] = i\epsilon_{ijk}J_k, \quad [P_i, J_j] = i\epsilon_{ijk}P_k \ . \tag{1}$$

For a particle, $\vec{J}$ can be decomposed into orbital and spin angular momentum:

$$\vec{J} = \vec{L} + \vec{S} \qquad \text{where} \qquad \vec{L} = \vec{X} \times \vec{P} \tag{2}$$

Since the P_i's commute, one can measure completely the momentum. The operators commuting with P_i are obviously the functions of $\vec{J}^2$ and $\vec{J}\cdot\vec{P} = \vec{P}\cdot\vec{J}$ (true equality, although $\vec{J}$ and $\vec{P}$ do not commute). This last operator is a pseudo-scalar; it is the <u>helicity</u>. Since

$$\vec{L}\cdot\vec{P} = \vec{P}\cdot\vec{L} = 0 \tag{3}$$

the helicity operator depends only on spin, $\vec{J}\cdot\vec{P} = \vec{S}\cdot\vec{P}$, and it is a good candidate for the polarization operator. The only trouble is that in special relativity, helicity

ISSN: 0094-243X/78/147/$1.50

cannot be defined for one particle with non-vanishing mass. (As we shall see, one needs two particle states to extend the helicity concept in special relativity to all particles).

In special relativity energy and momentum are the time component, P°, and space components, P^i, of a four vector, P^λ. Similarly J^k is the space part $J^k = M^{ij} = - M^{ji}$ (ijk = circular permutation of 1,2,3) of a skew symmetric tensor $M^{\mu\nu} = -M^{\nu\mu}$. The P^λ and $M^{\mu\nu}$ are the generators of the Poincare group and any function of them is a kinematical observable of special relativity. The P^λ commute

$$[P^\lambda, P^\mu] = 0 \qquad (4)$$

but the P^λ and the $M^{\mu\nu}$ among themselves do not commute. Moreover the separation of the total angular momentum $M^{\mu\nu}$ into an orbital and a spin part is not easy because a particle at rest for one observer is in motion for another observer and may have an orbital angular momentum for the latter but not for the former. However one must not confuse spin and polarization; they are two different concepts which coincide only in the rest system of a non-zero mass particle. The relativistic spin operator, part of the skew symmetric tensor $M^{\mu\nu}$, does not commute with the P^λ's. The polarization operator W does.

Let $\tilde{M}$ be the dual tensor of M:

$$\tilde{M}_{\lambda\mu} = \frac{1}{2}\, \epsilon_{\lambda\mu\nu\rho} M^{\nu\rho} \quad . \qquad (5)$$

The polarization operator is the axial vector

$$\underline{W} = \underline{\underline{\tilde{M}}} \cdot \underline{P} \qquad \text{i.e. } W_\lambda = \tilde{M}_{\lambda\mu} P^\mu \quad . \qquad (6)$$

It can be easily shown that it does commute with the P^λ's :

$$[P^\lambda, W^\mu] = 0 \quad . \tag{7}$$

Moreover, the only operators which commute with all P^λ's and $M^{\mu\nu}$ are $\underline{P}^2 = P^\lambda P_\lambda$ and $\underline{W}^2 = W^\mu W_\mu$. On the Hilbert space of single particle states (i.e. for an irreducible unitary representation of the Poincare group) these operators are multiples of the identity operator:

$$\underline{P}^2 = m^2 I, \quad \underline{W}^2 = -m^2 s(s+1) I. \tag{8}$$

where m and s are the mass and spin of the particle.

From the antisymmetry of $\tilde{\underline{\underline{M}}}$, we have that $\underline{P}\cdot\tilde{\underline{\underline{M}}}\cdot\underline{P}=0$, thus

$$\underline{P}\cdot\underline{W} = 0 \tag{9}$$

Note that the W_μ do not commute among each other in general. Indeed

$$[W_\mu, W_\nu] = i\epsilon_{\mu\nu\rho\sigma} P^\rho W^\sigma \quad . \tag{10}$$

Note also that $\underline{P}$ and $\underline{W}$ chosen here have the same dimension; this was done to deal with both $m = 0$ and $m \neq 0$ cases.

1. <u>Case m = 0</u>

This implies that $\underline{P}^2 = \underline{W}^2 = 0$ and $\underline{P}\cdot\underline{W}$ is always zero. It can be shown that two orthogonal vectors on the light cone must be colinear; therefore we can write

$$\underline{W} = \lambda\, \underline{P} \tag{11}$$

where the pseudo-scalar λ is the helicity of the mass zero particle. It is quantized: $\lambda = -\frac{1}{2}$ for neutrinos, $\lambda = +\frac{1}{2}$ for anti-neutrinos, and $\lambda = \pm 1$ for photons. We remark that in this $m = 0$ case the components of $\underline{W}$ commute among each other as equations (11) and (4) show.

2. <u>Case m ≠ 0</u>

In this case, a complete set of commuting observables is composed of the following components:

$$\text{four of } \underline{P},\ \text{one of } \underline{W},\ \text{and } \underline{W}^2 \tag{12}$$

Consider a particle of energy-momentum $\underline{p}$ with

$$\underline{p}^2 = m^2$$

(of course we have chosen the unit system in which $\hbar=c=1$.) Let $\underline{n}^{(o)} = \underline{p}/m$ and let $\underline{n}^{(i)}$ for i = 1,2,3 be together a set of four orthogonal four-vectors which obey

$$\underline{n}^{(\alpha)} \cdot \underline{n}^{(\beta)} = g^{\alpha\beta} \tag{13}$$

for $\alpha, \beta = 0,1,2,3$. We call such a set a <u>tetrad</u> of $\underline{p}$. Now we define the operators

$$S^{(\alpha)} = -\frac{1}{m} \underline{W} \cdot \underline{n}^{(\alpha)} = -\frac{1}{m} W^{\lambda} n_{\lambda}^{(\alpha)} \tag{14}$$

Equation (9) requires $S^{(o)}=0$. From (10), (13), (14) one finds that the three other operators $S^{(i)}$ satisfy:

$$[S^{(i)}, S^{(j)}] = i\epsilon_{ijk} S^{(k)} \ . \tag{15}$$

In the rest frame of the particle, the expectation values of these different operators are

$$\langle \underline{P} \rangle = (m, \vec{o}), \qquad \langle \frac{1}{m} \underline{W} \rangle = (o, \vec{w})$$
$$\text{with } w^k = \langle S^{(k)} \rangle = \langle J^k \rangle = \langle M^{ij} \rangle. \tag{16}$$

However one must not confuse these different operators. While for a particle at rest, $\frac{1}{m} W^k$, $S^{(k)}$ and $M^{ij} = J^k$ are the generators of the rotation group, their covariant meanings are quite different: the W^k are the components of <u>the polarization operator</u>, the $S^{(k)}$ are the generators of the little group of $\underline{p}$, i.e. the subgroup of the Lorentz group which leaves $\underline{p}$ invariant, while the $M^{\mu\nu}$ are the generators of the Lorentz group, and therefore the operators for the relativistic angular momentum components.

For a particle of spin s and mass $m \neq 0$, the polarization state is given by the expectation value of the irreducible symmetrized power of $\underline{W}$ up to degree 2s. Thus

$$W^{\lambda} = \frac{1}{m} \langle W^{\lambda} \rangle,$$

$$W^{\mu\nu} = \frac{1}{2m^2} \langle W^{\mu}W^{\nu} + W^{\nu}W^{\mu} - \frac{1}{2} g^{\mu\nu}\underline{W}^2 \rangle, \text{ etc...} \tag{17}$$

are respectively the dipole, quadrupole, etc...polarization tensors: Notice that $W^{\lambda\mu\nu\cdots}$ is completely symmetric in its indices, and that

$$p_{\lambda}W^{\lambda\mu\nu\cdots} = 0, \quad W^{\lambda}{}_{\lambda} = 0 \tag{18}$$

In the appendix, several applications of this covariant polarization formalism are outlined. Here we consider only the case of $m \neq 0$ for spin- $\frac{1}{2}$ particles. Such particles have only a dipole polarization. Their state is completely described by its energy-momentum $\underline{p}$ and its polarization $\underline{w}$. They satisfy

$$\underline{p}^2 = m^2; \; \underline{p}\cdot\underline{w} = 0; \; \underline{w}^2 = -(\text{degree of polarization})^2 \tag{19}$$

It is convenient to use the normalization $\underline{w} = \frac{2}{m} \langle \underline{W} \rangle$ so that the polarization degree is 1 for a completely polarized state and 0 for the unpolarized state.

In a given Lorentz frame fixed by a time axis such as the lab system or the center of mass system, one can choose a tetrad. Let us choose

$$\underline{n}^{(o)} = \underline{p}/m = (\gamma, \gamma\vec{v}); \quad \underline{n}^{(1)} = (o, \vec{s}_1)$$

$$\underline{n}^{(2)} = (o, \vec{s}_2); \quad \underline{n}^{(3)} = \underline{\ell} = (\gamma v, \gamma\vec{v}/v) \tag{20}$$

where $\gamma = E/m = (1-v^2)^{-1/2}$. Then we can define three pseudo-scalars ζ_i

$$\zeta_i = -\underline{w}\cdot\underline{n}^{(i)} \tag{21}$$

which implies

$$\underline{w} = \sum_{i=1}^{3} \zeta_i \underline{n}^{(i)}$$

Notice that ζ_3 is the longitudinal polarization. This of course is not a covariant concept. For instance in a $\pi^{\pm}$ decay in flight the $\mu^{\pm}$ is totally polarized; and this polarization is pure longitudinal in the π rest frame but not in the lab frame (see appendix). Of course $\underline{\ell}$ is the helicity axis. Since $\underline{P}\cdot\underline{W} = 0$, we can define the helicity, λ, in terms of operators as

$$\vec{P}\cdot\vec{W} = 2\lambda\ |W^\circ P^\circ| \tag{22}$$

This definition depends on the choice of a Lorentz frame (for $m \neq 0$). However we are not interested in isolated particles; we use them for making collisions. Given two particles with energy momenta $\underline{p}'$, and $\underline{p}''$ we can define the corresponding four-vectors $\underline{\ell}'$, $\underline{\ell}''$ by

$$\begin{aligned} \underline{\ell}' &= \frac{1}{\mathrm{sh}\chi}\,(\hat{\underline{p}}'\ \mathrm{ch}\chi - \hat{\underline{p}}'') \\ \underline{\ell}'' &= \frac{1}{\mathrm{sh}\chi}\,(\hat{\underline{p}}''\ \mathrm{ch}\chi - \hat{\underline{p}}') \end{aligned} \tag{23}$$

where

$$\hat{\underline{p}}' = \underline{p}'/m',\ \hat{\underline{p}}'' = \underline{p}''/m'',\ m^2 = (p'+p'')^2, \tag{23'}$$

and where

$$\mathrm{ch}\chi = \hat{\underline{p}}'\cdot\hat{\underline{p}}'' = \frac{m^2-m'^2-m''^2}{2m'm''},\quad \mathrm{sh}\chi = \frac{\sqrt{\Delta(m^2,m'^2,m''^2)}}{2m'm''} \tag{23''}$$

We define

$$\Delta(x,y,z) = x^2+y^2+z^2 - 2xy-2xz-2yz$$

By definition $\underline{\ell}'$ and $\underline{\ell}''$ are the longitudinal polarization or helicity quantization vectors of the particles in the rest system of $\underline{p} = \underline{p}'+\underline{p}''$. In a collision of particles, a and b , λ_a, λ_b are the s-channel helicity vectors.

In the collision process $a+b \to c+d$, we can define the t-channel helicity vectors. They are the longitudinal polarization vectors in the rest frame of $\underline{p}_a+\underline{p}_c$. This is usually called the Breit frame ($\vec{p}_a = -\vec{p}_c$).

Similarly one can define the u-channel helicity vectors.

Example: Consider a two body elastic collision in the center of mass in which a+b → c+d, where the beam particle a is scattered into the final particle c. The vector $\underline{n}$ is orthogonal to the scattering plane defined by $\underline{p}_a$, $\underline{p}_b$, $\underline{p}_c$, and $\underline{p}_d$, which are chosen according to the Basel convention, for which $\det(\underline{n}, \underline{p}_a, \underline{p}_b, \underline{p}_c) > 0$ This is common to the four tetrads p, ℓ, s, n. We have shown on the picture the space part of these four-vectors in the s-channel. Helicity reference frames use the ℓ vectors of the tetrad as the quantization axis. Transversity reference frames use n as the quantization axis, with $\underline{n} = \underline{n}^{(3)}$ for each tetrad in each channel.

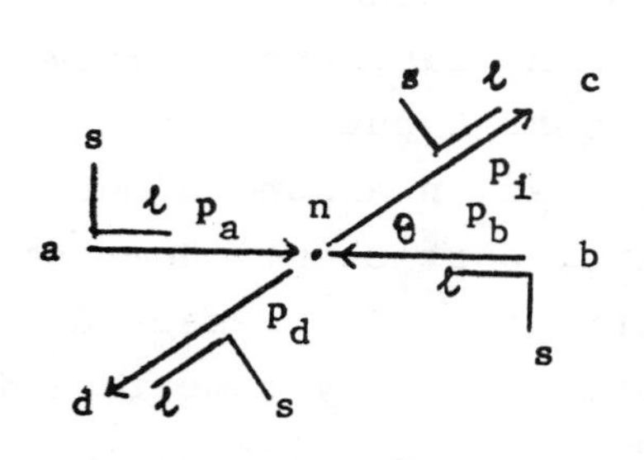

Figure 1:

To pass from the tetrad of one channel to that of another channel for a given particle, one has to do a rotation in the $\underline{\ell}$, $\underline{s}$ 2-plane about the crossing angle; this angle goes to zero in the forward direction limit.

If the scattering process a+b → c+d is time reversal invariant, then

$$A_{ij} = C_{ij} \tag{24}$$

i.e. initial and final state correlation parameters coincide.

The "Ann Arbor convention" advocates the s-channel tetrad in the c.m. system with the nice feature that the tetrads of the initial (or the final) particles are transformed into each other by a rotation R of π around $\vec{n}$. But for the study of the same reaction in the laboratory system ($\vec{p}_b = 0$) the advocated choice of tetrad might be natural for an experimental setting but it makes relation (24) <u>no longer valid</u> except for $A_{nn} = C_{nn}$. Indeed, if the advocated tetrad of the initial particles are obtained in the lab. system from those of the c.m. system by the boost Λ putting the b particle at rest, an additional rotation around $\vec{n}$ must be made for the tetrad of the final particles. However, the initial state correlation parameters A_{ij} measured in the c.m. and lab. frame could be directly compared if it were not for the very awkward choice of sign of $\vec{L}_b$ ("for historical reasons") the two initial tetrads are not transformed into each other by $\Lambda R \Lambda^{-1}$, the conjugation of the rotation of π. [Ed note: The convention has been changed to avoid this difficulty, which we decided was more important than historical reasons.]

Appendix. Other examples of application of the covariant polarization formalism.

1) $\pi \to \mu\nu$ decay. The polarization $\underline{w}$ of $\mu^{\pm}$ is orthogonal to $\underline{p}_\mu$ and depends only on $\underline{p}_\pi$, since $\underline{p}_\nu = \underline{p}_\pi - \underline{p}_\mu$. Using $\underline{p}_\nu^2 = 0$, one finds for $\mu^{\pm}$,

$$\underline{w} = \pm \frac{1}{\cos 2\omega} \left(\frac{\underline{p}_\pi}{m_\pi} \sin 2\omega - \frac{\underline{p}_\mu}{m_\mu} \right)$$

where $\frac{m_\mu}{m_\pi} = \mathrm{tg}\,\omega$, $\omega = .648\ldots$

2) Density matrix of a Dirac particle with non vanishing mass. The Dirac equation is

$$(\not{p} - m) u = 0.$$

The quantity

$$\rho(\underline{p}) = \frac{(1+\gamma^5 \not{w})(\not{p}+m)}{4m}$$

was introduced in L. Michel, A.S. Wightman, Phys. Rev. 98, (1955) 1190 where it is also extended to zero mass particles.

The covariant relation between the 2 x 2 density matrix formalism and Dirac matrices has been established for instance in C. Bouchiat, L. Michel, Nucl. Phys. 5, (1958) 416. If τ_k are the three Pauli matrices, with $\epsilon, \epsilon' = \pm 1$

$$u_\alpha(\underline{p}, \epsilon\underline{w})\bar{u}_\beta(\underline{p}, \epsilon'\underline{w}) = \frac{1}{4}[(\delta_{\epsilon\epsilon'} + \gamma^5 \not{n}^{(k)}(\tau_k)_{\epsilon'\epsilon})(\not{p}+m)]_{\alpha\beta}$$

where $\underline{n}^{(k)}$ for k = 1,2,3 form with $\underline{n}^{(o)} = \underline{p}/m$ the tetrad defined in (13).

3) Covariant density matrix of a particle with arbitrary spin. Let $s = \frac{n}{2}$ and $m \neq 0$. See L. Michel, N. Cim. Supl. 14 (1959) 95, where it was shown that

$$\rho(p) = \frac{1}{2s+1} w_{\lambda_1} \frac{W^{\lambda_1}}{m} + w_{\lambda_1\lambda_2} \frac{W^{\lambda_1}}{m}\frac{W^{\lambda_2}}{m} - \cdots$$

$$+ (-1)^n w_{\lambda_1\lambda_2\ldots\lambda_n} \frac{W^{\lambda_1}}{m}\frac{W^{\lambda_2}}{m} \cdots \frac{W^{\lambda_n}}{m}$$

Here the $w_{\lambda_1\lambda_2\ldots\lambda_k}$ are completly symmetric tensors, orthogonal to $\underline{p}$, and traceless; they are the covariant form of the polarization multipoles.

$$w_{\lambda_1\ldots\lambda_k} = w_{\lambda_{i_1}\ldots\lambda_{i_k}},$$

$$w^{\lambda}{}_{\lambda\,\lambda_3\ldots\lambda_k} = 0,$$

$$p^{\lambda} w_{\lambda\,\lambda_2\ldots\lambda_k} = 0.$$

The $w_{\lambda_1\ldots\lambda_k}$ are all even under parity since $\underline{W}$ is an axial vector. For instance a parity conserving two body decay of a spin-s particle measures only the even multipoles of the polarization when the polarization of the decay product is not observed. Indeed the observables are of the form $\mathrm{tr}\rho(\underline{p})A$, i.e. linear in $\rho(\underline{p})$ and Lorentz invariant. In the decay $p \to p_1 + p_2$ there is only one four-vector linearly independent from p, namely $q = p_1 - p_2$. When k is odd, $w_{\lambda_1 \ldots \lambda_k} q^{\lambda_1} \ldots q^{\lambda_k}$ is a pseudo-scalar.

<u>4) Polarization effects in Møller scattering</u>. In the Born approximation, which is good for small momentum transfer, $A^a = A^b = P^c = P^d = 0$. There is no analyzing power or polarizing power because it is a one photon exchange and all amplitudes are relatively real (in helicity).

All other effects can be computed in a completely covariant manner: cf C. Bouchiat, L. Michel, Compt Rend. Acad. Sci. Paris <u>243</u> (1956) 692, and Nucl. Phys. <u>5</u> (1958) 416; we also computed the spin effects for Bhabha scattering. We gave the results with arbitrary tetrads $\underline{n}^{(i)}$ for each particle, in terms of the invariants $\varkappa, \lambda, \mu$ traditionally used since the 1930's. We give the value of A_{ij} for Møller scattering in terms of the now fashionable $s=2m^2(1+\varkappa)$, $t=2m^2(1-\lambda)$, and $u=2m^2(1-\mu)$. These satisfy $s+t+u=4m^2$ and $\varkappa-\lambda-\mu=1$.

$$A_{ij} = N_{ij}/D$$

$$\begin{aligned} N_{ij} = & -\underline{n}_a^{(i)} \cdot \underline{n}_b^{(j)} 2(t^2u^2 - 4m^2tu(m^2-t-u)) \\ & -m^{-2}(\underline{n}_a^{(i)} \cdot \underline{p}_c)(\underline{n}_b^{(j)} \cdot \underline{p}_d) 4u^2(t+m^2) \\ & -m^{-2}(\underline{n}_a^{(i)} \cdot \underline{p}_d)(\underline{n}_b^{(j)} \cdot \underline{p}_c) 4t^2(u+m^2) \\ & -[(\underline{n}_a^{(i)} \cdot \underline{p}_c)(\underline{n}_b^{(j)} \cdot \underline{p}_c) + (\underline{n}_a^{(i)} \cdot \underline{p}_d)(\underline{n}_b^{(j)} \cdot \underline{p}_d)]4tu \end{aligned}$$

$$\begin{aligned} D = & \, t^2(2t^2+2ut+u^2-8m^2t+8m^4)+u^2(2u^2+2ut+t^2-8m^2u+8m^4) \\ & +2ut[(u+t)^2-4m^4] \end{aligned}$$

When we use the s-channel helicity tetrad defined in the text (see fig. 1; this is the Ann Arbor convention in the center of mass) in the limit $s/m^2 \gg 1$, we need only one parameter $\eta = tu/(s-4m^2) = (\sin\theta)^2/4$ to express

$$A_{\ell\ell} = \eta(2-\eta)/(1-\eta)^2$$

$$A_{ss} = A_{nn} = \eta^2/(1-\eta)^2$$

All other A_{ij} vanish for all s and t except $A_{\ell s}$ = $A_{s\ell}$ which decreases as $m/\sqrt{s}$ when $s/m^2 \to \infty$. Since $\sigma_{-\ell,\ell} = \sigma_{\ell,-\ell}$, we have $A_{\ell,\ell} = A_{-\ell,-\ell} = (\sigma_{\ell,\ell} - \sigma_{-\ell,\ell})/(\sigma_{\ell,\ell} + \sigma_{-\ell,\ell})$. This yields

$$\frac{\sigma_{\ell,\ell}}{\sigma_{-\ell,\ell}} = \frac{1+A_{\ell\ell}}{1-A_{\ell\ell}} = (1-4\eta+2\eta^2)^{-1}$$

which increases from 1 to 8 as θ goes from 0° to 90°. (This is used in F. Low's report). The cross section of electrons with the same helicity is always larger than that with opposite helicity. We also have

$$\frac{\sigma_{nn}}{\sigma_{-nn}} = \frac{\sigma_{ss}}{\sigma_{-ss}} = (1-2\eta+2\eta^2)/(1-\eta)$$

which increases from 1 to 5/4 in the same angular range. In the lab system, the A_{ij} have the same value except that the strange Ann Arbor convention requires that one change the sign of $A_{\ell\ell}$ and A_{ss}. [Ed. note: the convention was changed so that the signs of $A_{\ell\ell}$ and A_{ss} do not change in going from c.m. to lab frame.]

POSSIBLE EXPERIMENTS WITH POLARIZED NEUTRONS AT THE ZGS AND AT FNAL

Lawrence W. Jones
University of Michigan

I. INTRODUCTION

Experiments with polarized neutrons at high energy have interested me recently for at least five reasons as follows: (1) Our Michigan group has over the past decade evolved a program of hadronic physics with neutron beams at the Bevatron, the ZGS, the AGS, and at FNAL so that we feel quite familiar with the problems and techniques in this area[1]. (2) The Argonne ZGS has discussed plans to accelerate polarized deuterons; the neutrons from the stripped deuteron beam will provide a collimated, intense, reasonably monochromatic polarized neutron beam of up to 6 GeV. (3) The observed polarization of inclusively-produced $\Lambda°$s by Overseth's group at FNAL and CERN begs the investigation of polarization in other inclusive phenomena such as neutron production[2]. (4) J. Rosen et al. have recalled[3] that the Schwinger Effect[4] can produce up to 100% polarization of neutrons in elastic scattering at very small angles at high energies and outlined possible experiments at Fermilab. (5) During the winter and spring of this year I was a guest of Westfield College (London) where Elliot Leader stimulated my interest in various polarization phenomena accessible though small angle neutron scattering.[5] Drawing these factors together has lead to the conception of a series of possible experiments at both Argonne and Fermilab which I will sketch herein.

Fundamental to the class of experiments I am discussing is the interference between the electromagnetic

ISSN: 0094-243X/78/158/$1.50

(here magnetic) scattering and the nuclear scattering. As E. Leader has noted most succinctly[6], polarization effects are maximal where these amplitudes are equal at a value of transverse momentum $p_\perp^\circ$ in elastic scattering given by

$$p_\perp^\circ = \frac{4\pi\alpha Z\mu}{m\sigma_T(nA)}$$

for neutrons of mass m on a nucleus of charge Z and mass number A. For hydrogen and lead, $p_\perp^\circ \cong 1.8$ MeV/c, and for uranium, $p_\perp^\circ \cong 2.0$ MeV/c. It is convenient to define $q = p_\perp/p_\perp^\circ$. At very small angles, the magnetic scattering is proportional to $(-t)^{-1}$ while the nuclear term is nearly constant, so that the polarization varies with q as

$$P = \frac{2q}{1+q^2}$$

In Figure 1 we have plotted P vs. q and q^2 (proportional to -t). It is convenient that P falls only slowly for $q > q_o$ so that $\bar{P} \cong 0.7$ for $1 \le q \le 4$ and $\bar{P} \cong 0.4$ for $3 \le q \le 10$ (averaged over $p_\perp$, not t). In order to design an experiment, we want to consider the fraction of the incident flux which may be scattered into a range of t useful for polarization analysis. The relevant quantity is

$$\frac{1}{\sigma_T}\left(\frac{d\sigma_e}{dt}\right)\delta t .$$

A useful, conservative assumption is to consider a target 1/3 of an interaction mean free path thick. In this case, the fraction of incident flux scattered into certain ranges of δt from hydrogen and lead targets

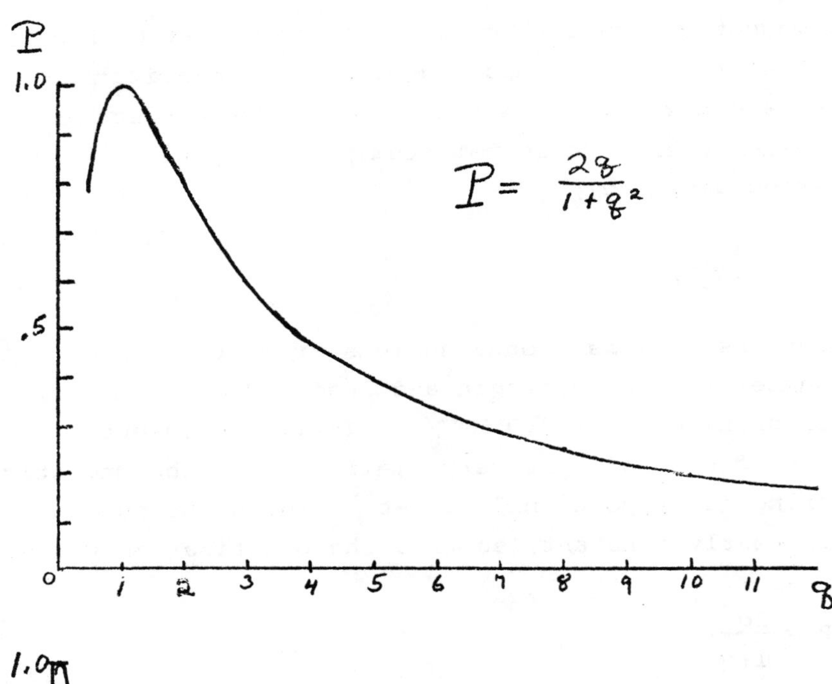

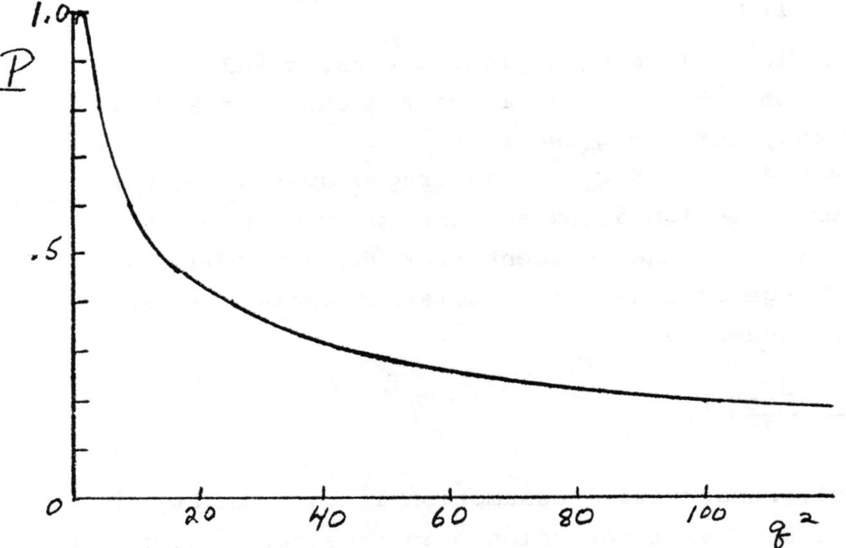

Figure 1. Plot of P vs. q and q^2 where $P = 2q/(1+q^2)$.

is indicated in Table I.

TABLE I

Fraction of incident flux elastically scattered into given ranges of q for 1/3 λ targets of H and Pb

q	A	$\bar{P}$	$1/3\left[\frac{1}{\sigma_T}\left(\frac{d\sigma_e}{dt}\right)\delta t\right]$
1-4	H	.7	2×10^{-5}
1-4	Pb	.7	2×10^{-3}
3-10	H	.4	1.2×10^{-4}
3-10	Pb	.4	1.2×10^{-2}

II. ZGS EXPERIMENTS

It is expected that the ZGS will produce a beam of polarized deuterons of about 5×10^9 per ZGS pulse at 12 GeV. From the analysis of W.T. Meyer[7] electromagnetic stripping occurs at this energy with a forward cross section $d\sigma/d\Omega \cong 10^{-21}$ cm^2/sr. In this case both the neutron and proton leave the stripping target, so that neutrons may be tagged by coincidence detection of the protons. Alternatively, nuclear stripping may be used wherein one nucleon is absorbed in the target and the other continues. This occurs with about ten times the cross section of EM stripping. As our proposed experiments do not require tagging it seems that the larger cross section is appropriate to use.

For a beam defined at an angle corresponding to q=1 at 6 GeV/c, $\theta \simeq (2\ \text{MeV/c})/(6\ \text{GeV/c}) = 3 \times 10^{-4}$ rad, so that $\delta\Omega \simeq 10^{-7}$ sr. (Strictly speaking, this θ could be the half-angle of the beam cone from a point source, however it is probably more realistic to assume a beam stripper target or neutron source of size comparable to the scattering target so that the angle subtended by the scattering target from each point at the source is 10^{-7} sr and the incident angles range over $\theta = \pm 3 \times 10^{-4}$ rad.) With a lead or uranium stripper of $1/3\lambda$ thickness a beam of 10^{-4} n per incident d would be contained in this solid angle, or about 5×10^{5} polarized neutrons per pulse. A possible experimental program at the ZGS might proceed as follows:

A. Verify the Schwinger Effect, develop polarimeter, and measure the neutron beam polarization. The beam as defined above and incident on a lead target of $\lambda/3$ length could then be detected by a neutron spatial detector in a plane downstream from the scatterer a distance equal to the distance between the stripping and scattering targets, as in Figure 2. A beam plug would be useful to block out the central (unscattered) beam; this could be a steel rod 4 or 5 feet in length (8 or 10 λ). If the incident deuteron beam spot on the stripping target were 1cm diameter and the scattering target the same, the two targets would be separated by about 30 m. The detector would then be 30 m beyond the scattering target, and the beam plug would be just over 2 cm diameter. A larger collimator of about 10 cm diameter concentric with the plug would be useful to keep down counting rates from events scattered at larger, less interesting angles. The neutron detector could be the same apparatus used in our group's experiments on np and nd elastic scattering at the ZGS and subsequently at FNAL.[8] This detector is an assembly of alternating magnetostrictive wire spark chambers and 0.5 inch Zn converter plates in addition to several trigger scintillators. The system records the tracks of secondaries from converted neutrons and, after extrapolating to the vertex,

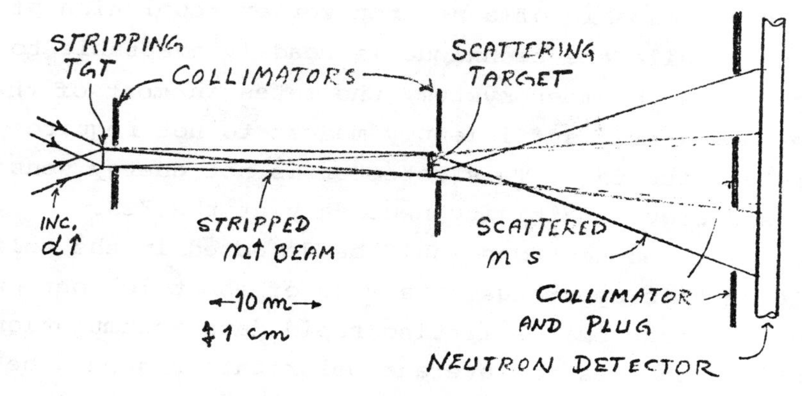

Figure 2. Experiment to study Schwinger Effect at the ZGS. Not shown are magnets to s eep charged particles and to rotate neutron polarization. Note 500:1 ratio of horizontal to vertical scales.

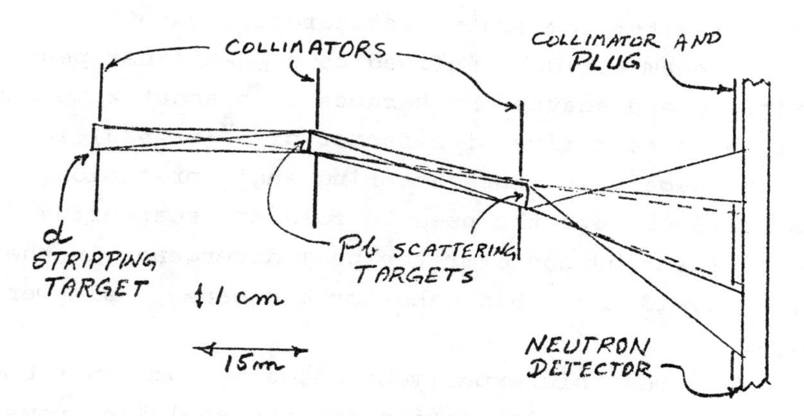

Figure 3. Double scattering experiment to unambiguously verify the Schwinger Effect. The first Pb scattering target produces polarized neutrons incident on the second lead scattering target.

was proven capable of a neutron vertex resolution of ≤ 0.1cm. While the technique is readily adaptable to a proportional chamber system, the rates in most of these experiments are sufficiently modest to not require a faster detector. This detector is not energy sensitive, nor is energy sensitivity necessary at the ZGS.

Scattered neutrons would be detected in the solid angle of this apparatus at a rate of about 10^3 per pulse (single scattered) permitting rapid data accumulation of polarization. As the average polarization should be 70%, even an hour of running would overkill the polarization determination. It would be necessary to check the centering of all beam elements (collimators, targets, etc.) and it would also be desirable to rotate the plane of polarization of the neutrons with magnets either before or after the scattering target, or both.

The beam could be defined to a much finer pencil (stripping and scattering targets both about 2 mm diameter) at a sacrifice of a factor of 5^4 (or 625) in rate but a gain in the scattering angle precision. This would obviate the need to fold the scattering angular distribution over the beam divergence in the data analysis and would make for a cleaner, simpler analysis.

Of course this experiment actually sums over both the neutron beam polarization and the analyzing power of the scattering process. To quantitatively and uniquely check the Schwinger Effect, it would be desireable to do a double scattering experiment. Here a two cm aperture centered at $q \cong 2$ in the plane of the collimator would redefine the now-polarized scattered beam; a second scatterer would replace the neutron detector, and the neutron detector would then be moved 30 m further down stream, again behind a beam plug and collimator (Figure 3). With 10^3 neutrons scattered into the plane of the second scatterer, only about 50 neutrons pass through the second collimator and hence only one neutron in ten pulses is scattered into the range $1 \leq q \leq 4$ by the second scatterer. Of course opening up the aperture of first and second collimators to accept twice the range of $p_\perp$ would increase the rate of detected neutrons by a factor of 16, with a corresponding loss of polarization.

B. The polarized neutrons could be scattered from an unpolarized hydrogen target, and the recoil neutron distribution in φ near $q = 1$ analyzed to determine ρ. This is not terribly attractive, as $P \propto (1+\rho^2)^{-\frac{1}{2}}$, and even $\rho = 0.3$ produces only a 5% decrease in P from its value (100%) when $\rho = 0$.

C. A more challenging experiment, and potentially a very interesting one, might be a double scattering experiment wherein a liquid hydrogen target is the first scatterer and a lead scatterer serves as an analyzer in a second scattering as in Figure 4. Here it would be necessary to open up the collimators to accept a beam solid angle corresponding to $\Delta q \sim 3$-5. Specifically, if the beam were collimated to 5 cm diameter at the hydrogen target and the scattered beam also to 5 cm at the Pb "analyzer", and again if each scatterer were $\lambda/3$ thick the rate would be about one double-scattered neutron per pulse. This experiment would permit one to measure the depolarization parameter D_{LS}, and hence ρ, directly.

The larger bite of q in each scatter would reduce the sensitivity of the measurement for the sake of

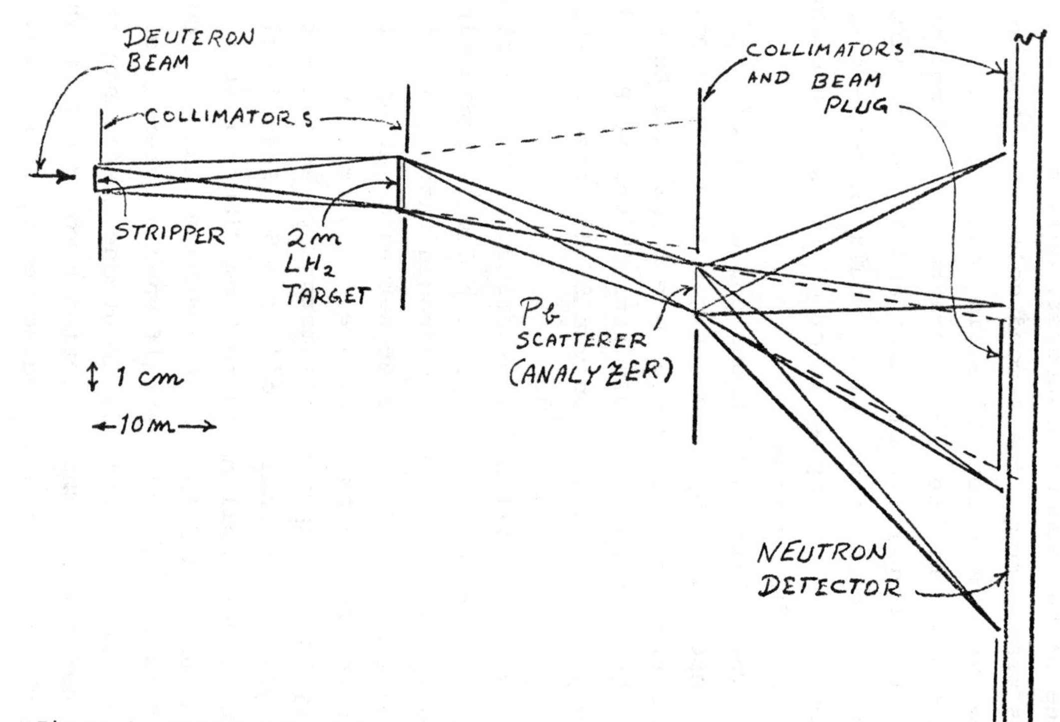

Figure 4. Geometry for determination of D_{SL} by double scattering on first liquid hydrogen and second lead.

counting rate, and an optimum found which would maximize the precision in determination of a physical parameter (e.g. ρ) in a given data collection time.

D. Scattering from a polarized target. Even a long polarized target contains only about 1g cm^{-2} of hydrogen, so that the fraction of the incident flux scattered into $1 \leq q \leq 4$ by protons in a polarized target would be about one per pulse with an incident beam of 5×10^5 polarized neutrons per pulse. Again, rates could be improved by opening collimators at the expense of reduced polarization sensitivity. This type of measurement would permit direct determination of A_{SL} and A_{LS}, and hence $\mathrm{Re}(\Phi_1 - \Phi_3)$ and $\mathrm{Re}\ \Phi_2$.

E. Double scattering from a polarized target. This is sheer blue sky, wherein a polarized target would be substituted for the liquid hydrogen target of B(above). The rates would be very low indeed; even with gimmicks it would be hard to reach one detected event per pulse.

III. FERMILAB EXPERIMENTS

We have routinely employed a neutron beam at Fermilab (beam M3 in the Meson Lab) containing a flux from 10^5 to 10^7 per pulse in a solid angle between 10^{-8} and 10^{-10}sr[9]. This beam is nominally unpolarized; its polarization has never been explored, or even questioned. However it is produced by a proton beam incident on a Be target at angles of 0.5 to 1.75 mr (variable), and recent developments have permitted larger targeting angles. It seems likely that neutrons are polarized to the extent that Λ°s have been found to be, e.g. $P \simeq 0.15p_\perp$ for $0.4 < x < 0.7$. If so, 300 GeV neutrons ($p_\perp \sim .15-.5$ GeV/c) might be polarized by up to several percent.

This polarization is a major uncertainty in the design of Fermilab experiments.

The Fermilab neutron beam is a "white" spectrum so that an energy-sensitive detector would be necessary. Our group has routinely used an ionization calorimeter with a 12% FWHM resolution at 300 GeV.[10] We have not so far combined this with a vertex detector (chamber) although this should be straightforward. At least several experiments suggest themselves.

A. Study of beam polarization. The beam polarization could be analyzed by employing the Schwinger Effect directly, using a 1/3 λ lead scatterer and a configuration as sketched in Figure 2. At Fermilab the beam target is 1/16 inch and may be collimated to 1/16 inch at 600 ft corresponding to $\theta = \pm 1.7 \times 10^{-5}$. The detector at 1250 feet (under the mezzanine) could then explore the scattered beam asymmetry. As at the ZGS magnets could be used to rotate the neutron beam polarization in order to cancel some systematic errors. (It is fortunate here that the spin precession is independent of momentum.) There would be 10^2 - 10^3 neutrons scattered per pulse into $1 \leqslant q \leqslant 4$ near the peak of the neutron spectrum in this experiment. I have assumed in this discussion that the Schwinger Effect and scattering in a lead analyzer would have already been studied and verified in experiments at the ZGS, so that a double scattering experiment would not be necessary.

If polarization of the inclusively-produced neutron beam is found, it would be of obvious interest to explore it quantitatively, as functions of target as well as neutron x and $p_\perp$. This experimental program would be analogous to the studies of $\Lambda°$ polarization in inclusive production.

B. Production of a polarized beam and study of $\Delta\sigma$

If the beam polarization were 10% or greater, it could be used as a polarized beam directly to study such quantities as $\Delta\sigma_L$ and $\Delta\sigma_S$ using a polarized proton target and a simple transmission experiment. If $\Delta\sigma/\sigma \cong 1\%$ and P(beam) = 10% the 1 g cm^{-2} of the polarized target would lead to a 2.4×10^{-5} change in transmission when the target or beam polarization is reversed. The systematic uncertainties in seeing such a small effect may be difficult but not impossible. If the beam polarization is too small, a polarized beam could be achieved by scattering on lead. With $q \simeq 3.$ a beam flux of the order of 10^2 to 10^3 per pulse could be achieved, with $P \geqslant 50\%$. This flux is still sufficient to study $\Delta\sigma_L$ and $\Delta\sigma_S$ as long as the systematic effects can be controlled.

C. Double scattering experiments.

Experiments to determine D_{SL} as under IIC above could be carried out at Fermilab. Here a long hydrogen target would be used to scatter the neutron beam and the polarization of the scattered neutrons could be determined from a second scattering on lead. Either the incident beam would be sufficiently polarized, in which case the numbers are analogous to the ZGS case (with the incident beam higher and incident polarization smaller by about an order of magnitude) or it would be necessary to effect an initial scattering on lead to polarize the beam, with a corresponding sacrifice in intensity and necessitating opening the various apertures as discussed earlier. The overall

M3 beam length, over 2000 ft, seems sufficient for the triple scattering.

D. Scattering on a polarized target

Again the arguments and numbers as in IID for the ZGS case, and as in IIIC above, the same caveats concerning the incident beam intensity and polarization pertain.

E. Other experiments

Double scattering on a polarized target (as in IIE) is an obvious but difficult possibility. It is also clear that the polarization in "normal" ($-t \geqslant 0.1(\mathrm{GeV}/c)^2$) elastic scattering could be studied. These and further studies made possible given a polarized neutron beam are obvious interesting extensions of an experimental program.

TABLE II

Physics Experiments with Polarized Neutrons

	Experiment	Reaction[a]	Determine[b]
I.	ZGS		
	A. Schwinger Effect	$n\uparrow Pb \rightarrow n(\theta,\varphi)Pb$	
	B. Hydrogen Scattering	$n\uparrow p \rightarrow n(\theta,\varphi)p$	ρ
	C. Double Scattering	$n\uparrow p \rightarrow n\uparrow\downarrow(\theta,\varphi)p$	D_{SL}, ρ
	D. Scattering on a Polarized Target	$n\uparrow p\uparrow\downarrow \rightarrow n(\theta,\varphi)p$	A_{SL}, A_{LS}
	E. Double Scattering on a Polarized Target	$n\uparrow p\uparrow\downarrow \rightarrow n\uparrow\downarrow(\theta,\varphi)p$	
II.	FNAL		
	A. Analyze Beam	$pBe \rightarrow n\uparrow(\theta,\varphi)X$ $n\uparrow Pb \rightarrow n(\theta,\varphi)Pb$	
	B. Cross Section Beam	$n\uparrow p\uparrow\downarrow \rightarrow$ anything	$\Delta\sigma_L$ $\Delta\sigma_S$
	C. Double Scattering	$n\uparrow p \rightarrow n\uparrow\downarrow(\theta,\varphi)p$	D_{SL}, ρ
	D. Scattering on a Polarized Target	$n\uparrow p\uparrow\downarrow \rightarrow n(\theta,\varphi)p$	A_{SL}, A_{LS}

---

a. The arrows represent known polarization, longitudinal or transverse, of the preceeding particle.

b. The notation follows E. Leader, refs. 5 and 6.

References

1. M.J. Longo *et al*., Particles and Fields, 1975, Proc. of Seattle Conf., (Am.Inst. Phys.) 412 (1975).
2. G. Bunce, *et al*., Phys. Rev. Lett. 36, 1113 (1976) K. Heller, *et al*., Phys. Lett. 68B, 480 (1977).
3. J. Rosen, "Design of a High Energy Polarized Neutron Beam for NAL" supplement to proposal for E-27 (unpublished).
4. J. Schwinger, Phys. Rev. 73, 407 (1978).
5. N.H. Buttimore, E. Gotsman, and E. Leader, "Spin Dependent Phenomena Induced by Electromagnetic-Hadronic Interference at High Energies" Westfield College, London, (unpublished, 1977).
6. E. Leader, "Usefulness of Spin-Dependent Electro-magnetic Hadronic Interference Experiments" TH.2386-CERN (unpublished, 1977). We have adopted the notation of this paper.
7. W.T. Meyer, "A Monoenergetic Polarized Neutron Beam at the ZGS", ANL/HEP 7441 (unpublished, 1974).
8. J. Stone, *et al*., Phys. Rev. Lett. 38, 1315 (1977). J. Stone, "A Detector for High Energy Neutrons with Good Spatial Resolution and High Efficiency", UM HE 76-12 (unpublished, 1976).
9. M.J. Longo, *et al*., "Characteristics of the Neutral Beam at NAL" UM HE 74-18 (unpublished,1974)
10. L.W. Jones, *et al*., Nucl. Instr. and Meth. 118, 431 (1974).

ONE PHOTON EXCHANGE PROCESSES AND THE CALIBRATION OF POLARIZATION OF HIGH ENERGY PROTONS*

B. Margolis
McGill University, Montreal, Quebec

G. H. Thomas
Argonne National Laboratory, Argonne, IL 60439

ABSTRACT

We examine polarization phenomena in small momentum transfer high energy one-photon exchange processes in the reaction $p + A \to X + A$ where A is a complex nucleus and X is anything. We show that these polarizations can be related directly to photoproduction polarization effects in the reaction $\gamma + p \to X$ at low energies. We write explicit formulae for polarization effects in the case where $X \to \pi^\circ + p$.

INTRODUCTION

It has been suggested[1] that the polarization of high energy proton beams may be measured utilizing photo-production data at low energies, for example in the region of energies where one produces the Δ resonance. This is possible since the process

$$p + A \to X + A \tag{1}$$

initiated by one photon exchange, the Primakoff effect,[2] has a strong forward peak and actually can become dominant over strong interactions for a heavy target nucleus A at small momentum transfers. X is anything that can be made by one photon exchange. Given that X is the Δ resonance the dominance of one photon exchange at small angles has been established experimentally.[3]

We will show below that the process (1) carries identical polarization information to that obtained from the reaction

$$\gamma + p \to X . \tag{2}$$

For example, if the target proton is polarized the decay

$$X \to a + b \tag{3}$$

will show typical azimuthal asymmetry and this is identical to the asymmetry that will be found in the decay of X in reaction (1) viewed in the Jackson frame if the incident high energy proton is polarized.

*Work supported in part by The National Research Council of Canada, The Quebec Department of Education and The United States Department of Energy.

ISSN: 0094-243X/78/173/$1.50

DERIVATION OF RESULTS

Figure 1 shows the Feynman graph for the one-photon exchange process

$$p + A \to X + A, \quad X \to a + b \tag{4}$$

The amplitude for this process may be written in the form

$$A = (Q_1 + Q_2)^\mu \, M_\mu \, \frac{Ze}{k^2} \, F(k^2) \tag{5}$$

where M_μ is the current carried by the upper line $p \to X \to a + b$, and $F(k^2)$ is the form factor for the nucleus. It is advantageous for us to work in the rest frame of X in order to relate our results to photoproduction of X on polarized proton targets.

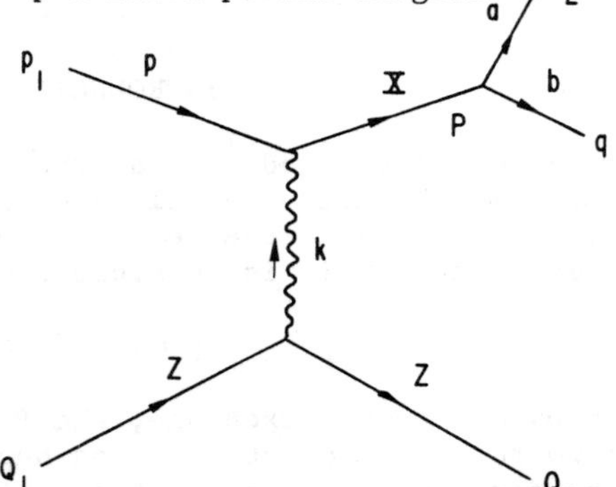

Fig.1. One photon exchange for process (4) with the kinematical variables indicated. The invariants are $Q_1^2 = Q_2^2 = M^2$, $p_1^2 = m^2$, $p_2^2 = m_2^2$, $P^2 = s_1$, $q^2 = \mu^2$, $k^2 = (Q_1 - Q_2)^2$, and $s = (p_1 + Q_1)^2$.

It follows from gauge invariance that

$$k^\mu M_\mu = 0 \; . \tag{6}$$

Choosing a coordinate system in the rest frame of X as in Figure 2 we then find that

$$A = 2Q_1^\mu \, M_\mu \, \frac{Ze}{k^2} \, F(k^2)$$

$$\approx - \frac{Ze}{k^2} \, F(k^2) \, \frac{2Mp_L}{\sqrt{s_1}} \left\{ \frac{\sqrt{s_1}}{m\nu} \, \underset{\sim}{P}_\perp \cdot \underset{\sim}{M}_\perp - \left(\frac{s_1}{p_L^2} + \frac{s_1 P_\perp^2}{m^2 \nu^2} \right) M_\parallel \right\} \tag{7}$$

$$\approx -\frac{Ze}{k^2} F(k^2) \frac{2Mp_L}{m\nu} \underset{\sim}{P}_\perp \cdot \underset{\sim}{M}_\perp$$

under the condition

$$s_1/p_L^2 << \frac{\sqrt{s_1}\ P_\perp}{m\nu} . \qquad (8)$$

The quantity ν is the photon energy in the frame in which p_1 is at rest, $P_\perp$ is the transverse component of $\vec{P}$, the 3-momentum of X in the c.m. frame of the reaction (4), p_L is the proton laboratory momentum, m is the proton mass and s_1 the mass-squared of X.

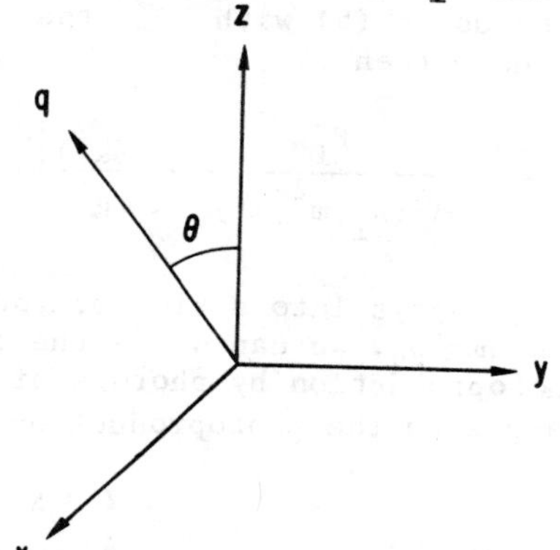

Fig.2. Coordinate system in the rest frame of X.

We can then write the differential cross section for the reaction (4) as

$$\frac{d\sigma}{ds_1 dk^2 dt} = \frac{1}{2\lambda^{\frac{1}{2}}(s,m^2,M^2)} \frac{1}{(2\pi)^5} \int d^4q\, d^4p_2 d^4Q_2\, \delta(q^2-\mu^2)\, \delta(p_2^2-m_2^2)$$

$$\delta(Q_2^2-M^2)\, \delta\left[s_1-(p_2+q)^2\right] \delta\left[t-(p_2-p_1)^2\right] \delta\left[k^2-(Q_1-Q_2)^2\right] \qquad (9)$$

$$\delta^4(p_2+Q_2+q-p_1-Q_1)\ |A|^2$$

where $\lambda(x,y,z)=x^2+y^2+z^2-2xy-2xz-2yz$. One then finds

$$\frac{d\sigma}{ds_1\ dk^2\ dt\ d\phi} = \frac{\pi}{16\lambda(s,m^2,M^2)\lambda^{\frac{1}{2}}(s_1,k^2,m^2)} \frac{1}{(2\pi)^5} |A|^2 , \qquad (10)$$

Here ϕ is the azimuthal angle of particle a or b in the frame in which X is at rest measured say relative to the polarization vector of the incident photon having momentum k as in Figure 3. As seen in (7), the photon is indeed polarized with polarization tensor $\hat{\varepsilon}_T = \hat{P}_\perp$ given by the direction of $\underset{\sim}{P}_\perp$. We can write (10) in terms of the cross section for the photoproduction of X,

$$\gamma + p \rightarrow X \rightarrow a + b \tag{11}$$

since this is given by

$$\frac{d\sigma(\gamma_\uparrow p_\uparrow \rightarrow ab)}{dt d\phi} = \frac{1}{32\pi^2} \cdot \frac{1}{\lambda(s_1, 0, m^2)} \left| \hat{\varepsilon}_T \cdot M_T \right|^2 \tag{12}$$

The current M^μ being as in (5) with $\hat{\varepsilon}_T$ the polarization of the incident photon. We have then

$$\frac{d\sigma}{ds_1 dk^2 dt d\phi} = \frac{Z^2\alpha}{\pi} \frac{P_\perp^2}{\lambda^{\frac{1}{2}}(s_1, m^2, 0)} \frac{|F(k^2)|^2}{k^4} \frac{d\sigma(\gamma_\uparrow p_\uparrow \rightarrow ab)}{dt\, d\phi} \tag{13}$$

For the case where X decays into a pion of four momentum q and a nucleon of four momentum p_2, we can write the c.m. differential cross section for photoproduction by photons of momentum k, polarization $p_\gamma = \varepsilon$ at an angle to the photoproduction plane ϕ, incident on

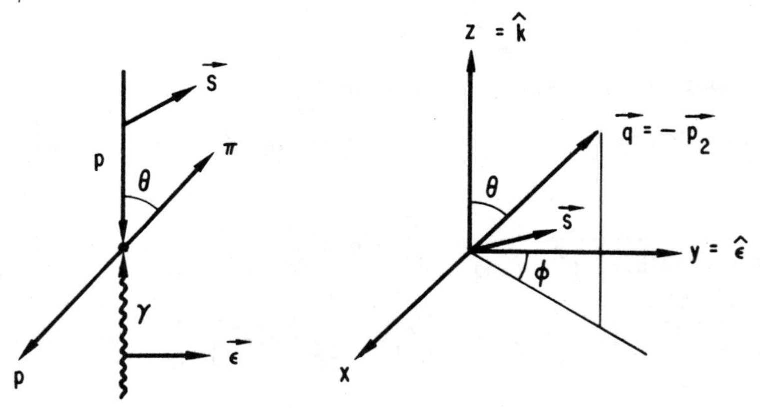

Fig. 3. Coordinate system in rest frame of X showing definition of ϕ.

a nucleon of four momentum p_1 and polarization $\vec{s}$ (see Figure 3) as[4]

$$d\sigma(\gamma_\uparrow p_\uparrow \rightarrow \pi^\circ p) = d\sigma(\theta) \Big\{ 1 - \Sigma(\theta) p_\gamma \cos 2\phi + T(\theta) s_y - H(\theta) s_x p_\gamma \sin 2\phi$$
$$- P(\theta) s_y p_\gamma \cos 2\phi + G(\theta) s_z p_\gamma \sin 2\phi \Big\} . \tag{14}$$

Here $d\sigma(\theta)$ is the unpolarized photoproduction cross section, the asymmetries $\Sigma(\theta)$ and $T(\theta)$ are the single polarization coefficients. The polarization of the recoil nucleon $P(\theta)$, $H(\theta)$ and $G(\theta)$ occur in (14) as double polarization coefficients. Expressions for the above quantities are as follows.[4] Given H_1, H_2, H_3 and H_4 the complex s-channel helicity amplitudes in the notation of Walker[5]

$$d\sigma(\theta) = \frac{1}{2}\frac{q}{k}\left\{|H_1|^2 + |H_2|^2 + |H_3|^2 + |H_4|^2\right\} \tag{15}$$

$$P(\theta) = -\frac{q}{k}\frac{1}{d\sigma(\theta)}\,\mathrm{Im}(H_1H_3^* + H_2H_4^*) \tag{16}$$

$$\Sigma(\theta) = \frac{d\sigma_\perp - d\sigma_\parallel}{d\sigma_\perp + d\sigma_\parallel} = \frac{q}{k}\frac{1}{d\sigma(\theta)}\,\mathrm{Re}(H_1H_4^* + H_2H_3^*) \tag{17}$$

where $d\sigma_\perp$ and $d\sigma_\parallel$ are the cross sections for complete polarization perpendicular and parallel to the photoproduction plane, the target proton being unpolarized.

Given target nucleons polarized in the direction perpendicular to the plane of production and unpolarized photons

$$T(\theta) = \frac{d\sigma^\uparrow - d\sigma^\downarrow}{d\sigma^\uparrow + d\sigma^\downarrow} = \frac{q}{k}\frac{1}{d\sigma(\theta)}\,\mathrm{Im}(H_1H_2^* + H_3H_4^*) \tag{18}$$

where $d\sigma^\uparrow$ and $d\sigma^\downarrow$ are the cross sections for meson production from nucleons polarized up and down in the direction $\vec{k}\times\vec{q}$. Finally,

$$H(\theta) = -\frac{q}{k}\frac{1}{d\sigma(\theta)}\,\mathrm{Im}(H_1H_3^* - H_2H_4^*) \tag{19}$$

$$G(\theta) = -\frac{q}{k}\frac{1}{d\sigma(\theta)}\,\mathrm{Im}(H_1H_4^* - H_2H_3^*) \tag{20}$$

REFERENCES

1. "A Method of Measuring the Polarization of High Momentum Proton Beams", D. G. Underwood, ANL-HEP-PR-77-56.
2. H. Primakoff, Phys. Rev. 81, 899 (1951).
3. W. Mollet et al., Phys. Rev. Lett. 39, 1646 (1977).
4. H. M. Fischer, "Photoproduction Experiments in the Nucleon Resonance Region", Proc. of the 1975 International Symposium on Lepton and Photon Interactions at High Energies, p.413-417.
5. R. L. Walker, Phys. Rev. 182, 1729 (1969).

ISOSPIN BOUNDS ON πN AND NN ELASTIC POLARIZATIONS AT HIGH ENERGIES

C. BOURRELY and J. SOFFER
Centre de Physique Théorique - CNRS
31, Ch. J. Aiguier - 13274 MARSEILLE CEDEX 2 (France)

L. MICHEL
I.H.E.S. - 35, Route de Chartres
91440 BURES-SUR-YVETTE (France)

Recent experimental results have been obtained on πN and NN elastic polarizations both at FNAL and CERN. As it is well known, isospin invariance implies some restrictions on these polarizations and we feel that these old results should be reconsidered quantitatively, in particular, at the highest available energies.

For πN scattering a classical result [1] is

$$|P_+ - P_-| < 2 \sin\omega \tag{1}$$

where P_+ and P_- denote respectively the $\pi^+ p$ and $\pi^- p$ elastic polarizations. The three unpolarized differential cross sections σ_+, σ_-, σ_o which correspond to $\pi^+ p$, $\pi^- p$ and the charge exchange reaction $\pi^- p \rightarrow \pi^o n$ must be such that $\sqrt{\sigma_+}$, $\sqrt{\sigma_-}$ and $\sqrt{2\sigma_o}$ are the three sides of a triangle and ω is the angle between the two sides $\sqrt{\sigma_+}$ and $\sqrt{\sigma_-}$, i.e. $\cos\omega = (\sigma_+ + \sigma_- - 2\sigma_o)/2\sqrt{\sigma_+\sigma_-}$. Using the experimental fact that σ_+ and σ_- are much larger than σ_o it is obvious that

$$\sin\omega < \sin\omega_{max} = \sqrt{\frac{2\sigma_o}{\sigma_s}} \tag{2}$$

where $\sigma_s = \sup(\sigma_+, \sigma_-)$. Then the bounds in (1) become

$$|P_+ - P_-| < 2\sqrt{2}\sqrt{\frac{\sigma_o}{\sigma_s}} \tag{3}$$

valid at all energies and scattering angles.

ISSN: 0094-243X/78/178/$1.50

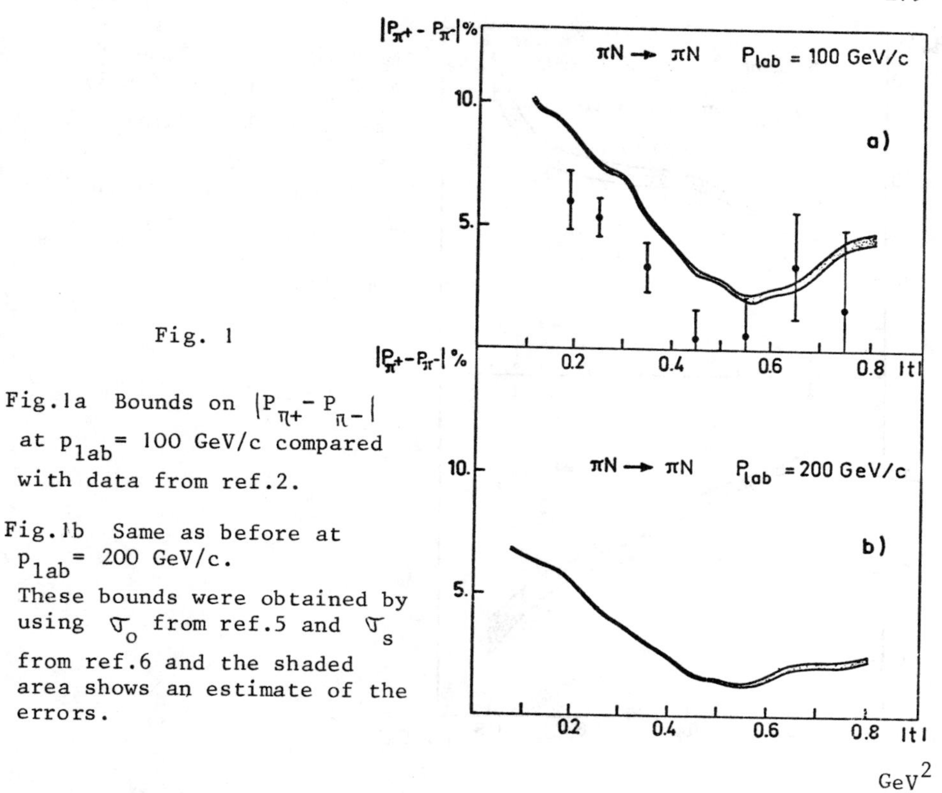

Fig. 1

Fig.1a Bounds on $|P_{\pi+} - P_{\pi-}|$ at p_{lab} = 100 GeV/c compared with data from ref.2.

Fig.1b Same as before at p_{lab} = 200 GeV/c.
These bounds were obtained by using σ_o from ref.5 and σ_s from ref.6 and the shaded area shows an estimate of the errors.

The difference of the two elastic polarizations can be bounded knowing only the unpolarized cross sections. Moreover the size of the bound is controlled by the rapid decrease with energy of σ_o and is minimum at the position of the dip of σ_o ($|t| \sim 0.6$ GeV^2). The results are shown in fig. 1 at p_{lab} = 100 GeV/c and 200 GeV/c. In fig.1a, we have compared the bound with recent data [2] which satisfy the constraint except at $|t| = 0.65$ GeV^2. This simple check should be made for future high energy experiments and any violation in a sizeable t range of these bounds would clearly indicate the presence of a Coulomb-nuclear interference polarization [3].

For NN scattering the analog of (3) is

$$|P_{pp} - P_{pn}| < 2 \sqrt{\frac{\sigma_o}{\sigma_s}} \tag{4}$$

where P_{pp} and P_{pn} denote respectively the pp and pn elastic polarizations, σ_o is the charge exchange differential cross section

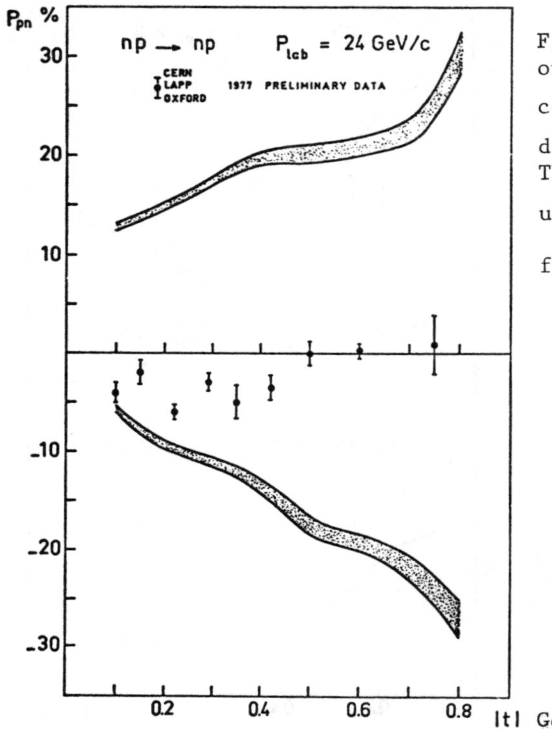

GeV^2

Fig.2. Upper and lower bounds on P_{pn} at p_{lab}= 24 GeV/c compared with preliminary data from ref.4.
These bounds were obtained by using σ_o from ref.7 and σ_s from ref.8.

Fig. 3. Bounds on $P_{pp}-P_{pn}$ at p_{lab}= 100 GeV/c and 280 GeV/c. These bounds were obtained by using σ_o from ref.9 and σ_s from ref. 10 .

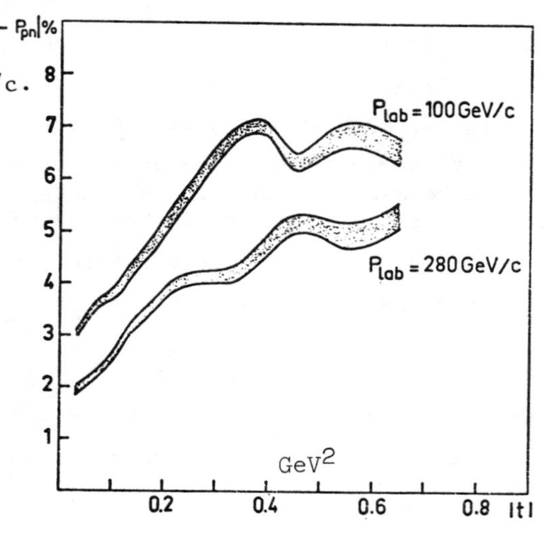

and $\sigma_s = \sup(\sigma_{pp}, \sigma_{pn})$. Knowing P_{pp} and the differential cross section this result gives the allowed domain for P_{pn}. This is shown at $p_{lab} = 24$ GeV/c in fig. 2, together with some P_{pn} preliminary data [4]. We have also calculated the bounds for the difference $|P_{pp} - P_{pn}|$ at $p_{lab} = 100$ GeV/c and 280 GeV/c, and the results are shown in fig. 3. A bound for P_{pn} will be obtained from these results when P_{pp} will be measured at these high energies.

REFERENCES

1. L. MICHEL, Brandeis Summer Institute, 1965, vol.1 (New York 1966).
   O. KAMEI and S. SASAKI, Nuovo Cimento 59, 534 (1969).
   G.V. DASS, J. FROYLAND, F. HALZEN, A. MARTIN, C. MICHAEL and S.M. ROY, Physics Letters 36B, 339 (1971).
2. I.P. AUER et al., Phys. Rev. Letters 39, 313 (1977).
3. C. BOURRELY and J. SOFFER, Nuovo Cimento Letters 19, 569 (1977).
4. CERN-LAPP-OXFORD Collaboration EXPT CERN-141 S.
5. A.V. BARNES et al., Phys. Rev. Letters 37, 76 (1976).
6. D.S. AYRES et al., Phys. Rev. D15, 3105 (1977).
7. M.N. KREISLER et al., Nuclear Physics B84, 3 (1975).
8. J.V. ALLABY et al., Nuclear Physics B52, 316 (1973).
9. H.R. BARTON et al., Phys. Rev. Letters 37, 1656 (1976).
10. M. KWAK et al., Phys. Letters 58B, 233 (1975).

DAY TO DAY MINI-SUMMARIES

The following pages contain the "mini-summaries" of each day's session by each group during the workshop. They were prepared on a day by day basis immediately after the end of each group discussion. By and large they are distilled from notes taken by the different session chairmen and from tape recordings made by the scientific secretary. They were not proofread for accuracy by the various chairmen, since speed was of the essence in getting these notes distributed to all participants at the earliest possible moment, usually in time for the next day's meetings.

Therefore it is inevitable that omissions and inaccuracies will exist in these notes. They are included in order that no ideas go unreported in these proceedings, and to give readers a sense of how the discussions went on a day to day basis. It must be emphasized, however, that we make no claims on their completeness, validity, or scientific accuracy.

A.J. Salthouse

ISSN: 0094-243X/78/182/$1.50

THEORY SESSION

October 19 Afternoon

F. Low and G. Thomas presiding

1. Asymmetries in $pp \rightarrow \Lambda + X$ at low $P_\perp$ in fragmentation processes. Generalize to any baryon. We want to understand the triple-Regge region.

2. Asymmetries in $pp \rightarrow \Lambda + X$ at large $P_\perp$ in hard processes.

3. Determine whether the asymmetry in $pA \rightarrow p\pi^\circ A$ can be calculated from the $\gamma p \rightarrow \pi^\circ p$ via the Primakoff effect.

4. ed scattering (both polarized). It is of theoretical interest?

5. Learn about the spin dependence of diffractive scattering.

6. Study the energy dependence in $pp \rightarrow \Delta^{++}n$; is this related to the spin dependence of diffraction? This may tie in to inclusive production of p and Λ.

7. What about polarization in CEX processes such as $\pi^- p \rightarrow \pi^\circ n$? This seems to be well studied already; little to be learned.

8. Can Coulomb interference be used in a polarimeter?

9. Extrapolation of LE pp amplitude analysis to HE, and predictions of asymmetries.

POLARIMETER WORKSHOP

October 19 Afternoon

J.B. Roberts presiding

Several kinds of polarimeters were discussed, using several approches

1) Coulomb scattering

   A. How good is the theoretical analysis? Can we calculate the Coulomb asymmetry at very small t?

   B. Perhaps this idea should be tested with 12 GeV protons at the ZGS

   C. Develop a polarimeter to use this method

   D. Can you measure deuteron polarization the same way? Problem: the polarization, which depends on the Coulomb-nuclear interference, is proportional to the magnetic form factor. This is 25 times smaller for deuterons than for protons, so asymmetry will be very small.

2) Use asymmetry in $pA \rightarrow p\pi^\circ A$ at small t

   A. Use Primakoff effect to connect $pA \rightarrow p\pi^\circ A$ at low t to $\gamma p \rightarrow \pi^\circ p$ at low energy which has large asymmetry. Hopefully this means $pA \rightarrow p\pi^\circ A$ will also have large asymmetry, but theoretical analysis is not straightforward.

   B. How complicated does the polarimeter have to be?

3) Use inclusive asymmetries, such as $pp \rightarrow \Lambda+X$

   A. There are asymmetries in $\pi^\pm$, π° between 8 and 24 GeV/c in unexplored kinematic regions.

   B. Suggest using the polarized target E61 at Fermilab and look in the fragmentation region for asymmetries in $\pi^\pm$, in the same kinematic region where asymmetries were seen at Argonne.

   C. This could be used as an easy polarimeter for either p or d beams.

LOW ENERGY COLLECTOR RINGS

October 19 Afternoon

H.F. Glavish presiding

Two kinds of rings were considered: a 500 KeV(pre-linac), and a 50 MeV (post-linac).

1) 500 KeV (protons)

A. Size 10-20m diam.

B. Estimated gain in intensity of about 10

C. Very small field; major expense is vacuum system

D. Assumed 100 μA source

E. Space charge limit in collector ring was taken as $3 \cdot 10^{10}$

F. Reached with 75 turn injection, taking 500 μsec

G. Single turn extraction from collector.

2) 50 MeV Ring (protons)

A. Space charge limit is higher, 10^{12}

B. Linac can only handle 10^{10}, so needs 100 pulses

C. Assume source of 100 μA, or $4 x 10^9$ p/6μs

D. 20 turn injection with inj eff ~ 1/2, capture eff ~ 1/2, linac eff ~ 1/2, so $20 \times 4.10^9 \times (1/2)^3 \sim 10^{10}$p are injected

E. Need electron cooling to fill collector

3) H^- polarized ions

A. Has advantage of stripping before injection into cyclotron

B. Limitations caused by multiple passes through stripping foil or gas; limits number of turns.

C. Could do 100-1000 turn injection

D. 1μA of H^- might be worth 10-100 μA of H^+ as far as loading a synchrotron is concerned. 5μA of H^- is available at present.

EXPERIMENT SESSION
October 20 Morning
O. Chamberlain presiding

1. Current theories strongly suggest that electromagnetic and weak interactions probe different properties than hadronic interactions. (Gluons carry no isospin, no charge, but color. Photons interact with electromagnetic currents. Leptons interact with the weak and electromagnetic currents.) Thus, we clearly need substantial programs with both proton and electron machines.

2. Inclusive reactions are more revealing of parton interactions than elastic scattering. (Analogy: Inelastic α-α scattering would be more revealing of proton-proton interactions than elastic α-α scattering.)

3. The inclusive Λ polarization observations of Bunce et al. show that there is strong spin dependence in inclusive processes. Presumably this reflects strong spin dependence in quark-quark interactions.

4. Some of the important future experiments then involve high-$P_\perp$ inclusive reactions, including their spin-dependent parts. This implies that we need high-intensity polarized proton beams, and superior polarized proton targets.

5. For a full spin analysis, $pp \to \Lambda X$ seems to be a most promising reaction, the lambda being self-analyzing as to its spin state.

6. To see directly quark-quark interactions, without being sensitive to the X (momentum fraction) distribution of the quarks within the nucleon, we need experiments at high-$P_\perp$ and at high energy.

7. There was some discussion of $NN \to \Delta N$. It involves quark spin-flip.

ACCELERATION OF POLARIZED PROTONS AND DEUTERONS

October 20 morning

E. Courant and L. Ratner

A survey was made of particular accelerators and the relative strengths of depolarizing effects at the ZGS; AGS; CERN PS; ISABELLE; FNAL; SPS; ISR.

The following conclusions were reached:

1) The intrinsic resonances at $8+\nu$, $16+\nu$ seem to behave in accord with calculations, within factors of 2.

2) There is reason to believe that computer models will agree with experimental results concerning imperfection resonances, if they have the right kinds of input.

3) Will use the computer models to compare the depolarization strengths at different accelerators for both types of resonances.

4) Reached a conclusion regarding relative strengths of resonances at AGS and ZGS: The AGS depolarizing intrinsic resonances are roughly twice as strong as at the ZGS, and the AGS imperfection resonances are about ten times stronger than at the ZGS, for comparable energies and orbit distortions.

5) It appears that tuning out depolarization at the AGS may be somewhat more difficult (but possible) than at the ZGS, below 20-25 GeV. Because of a very strong resonance at 25 GeV, tuning out the depolarization would be almost impossible at the AGS above 25 GeV.

6) A. Turrin submitted a report on the tune shift produced by quadrupole corrections, different from the Froissart-Stora formula.

7) Acceleration of polarized deuterons.

A. At the AGS no intrinsic and 2 imperfection resonances, which are weak.

B. At the ZGS, no intrinsic and 1 imperfection resonance, which is weak.

C. Therefore it will be much easier to tune out depolarizing resonances for deuterons at both the AGS and ZGS.

SOURCE WORKSHOP
October 20 Morning
H.F. Glavish presiding

Two kinds of source were discussed:

1) P. Chamberlin of LAMPF: Lamb shift source

A. Limitations associated with space charge effects and quenching.

B. No specific conclusions or ideas for improvement

C. This source has the ability to perform intrinsic polarization calibrations

D. It is relatively easy to stabilize the beam current by a feedback mechanism.

2) W. Kubischta of CERN: Velocity measurements of the CERN Atomic beam source

A. Cooled atomic beams can be generated

B. Supersonic flow is generated from the nozzle of the atomic beam source.

C. There was no specific explanation for the enhancement generated by pulsing

D. These velocity measurements are the first definitive ones for an atomic beam source, and they could form the basis for future design of dissociators.

COMBINED POLARIMETER AND HYPERON BEAM SESSION
October 20 Afternoon
G. Fidecaro reporting

A comparison was made between the CERN and FNAL hyperon projects (the FNAL project was presented by A. Yokosawa).

1. Motivation behind the FNAL experiment:
   A. Study of inclusive π^- production from polarized protons on polarized and unpolarized proton targets at large $P_\perp$ ($P_\perp$ = 6 GeV/c).
   B. Study of total cross sections ($\Delta\sigma_T$ and $\Delta\sigma_L$) aiming at a precision of ±10μb compared to $\Delta\sigma \sim$ 1000μb, in 2 days running. This seems to be possible at both FNAL and CERN.

2. FNAL beam design is based on 10^{13} protons on the primary target and a large acceptance, using superconducting quadrupoles. About 10^8 pppp expected. The resulting total deflection is zero both in the horiz and vert. plane. Momentum analysis is made in the vertical plane.

3. A combination of four bending magnets working in the two planes would permit one to rotate the spin ±90° in one plane. This scheme is already used at FNAL to depolarize the beam for polarization experiments.

4. There is no difficulty in principle in having similar intensity (10^8 pppp) at CERN and in using the same scheme to reverse the beam polarization.

5. Monitoring the polarization:
   A. Primakoff effect being considered at FNAL
   B. Polarization in the Coulomb-nuclear interference region is still under discussion in the experimental, theoretical, and polarimeter sessions. Discussions on this point will continue.

THEORY SESSION

October 20 and 21

F. Low and G. Thomas presiding

The theory people broke up into small groups who worked on the problems outlined on October 19.

1. Leader and Low worked on low t fragmentation in $pp \rightarrow$ baryon + X.

2. Soffer and Michel worked on high t fragmentation in $pp \rightarrow$ baryon + X.

3. Margolis and Thomas studied the $pA \rightarrow p\pi^{\circ}A$ asymmetry and the Primakoff effect.

4. ed - no work done

5. Thomas, Soffer, and Low studied the spin dependence of diffractive scattering

6. Soffer and Margolis looked at $pp \rightarrow \Delta^{++}n$ and spin dependence

7. CEX polarization - no work done

8. E. Leader will give a talk on polarization in Coulomb interference.

9. Leader and Thomas worked on extrapolation of LE amplitude analysis.

ACCELERATOR SESSION

October 21 morning

E. Courant and L. Ratner presiding

1. Acceleration of deuterons

   A. Expect that the intensity would be 2-5 times less than protons, for the same source. The source intensity would be unchanged, but the deuterons have lower overall transport efficiency in the accelerator.

   B. From an experimentalist's viewpoint, it is better to use the d beam as is, rather than strip into p beam, because resulting p beam would be more diffuse. Combined with the stripping efficiency, the proton intensity $\sim 10^{-3}$ deuteron intensity (crude estimate).

   C. No major difficulty foreseen in d acceleration. Deuterons have been successfully accelerated and stored at CERN, with ~30% of the proton intensity.

   D. The AGS linac is not presently suitable for d acc, and would only give 30 MeV as opposed to 200 MeV for p.

   E. The ZGS will be accelerating d in about a year, so more information will be available then.

2. Storage of protons at ISR and ISABELLE

   A. If an energy spread greater than 2% exists, it would straddle one of the imperfection resonances.

   B. (Since the factor for deuterons would be 25 times larger, d storage does not present any problems).

   C. Possibility of beam depolarization after long storage times: An experiment at the ZGS will hopefully improve existing knowledge of storage depolarization times. Present information is limited by systematic errors in previous experiments, which showed no depolarization. The errors were too large to set limits on storage times.

3. Accelerator configurations that preserve the longitudinal polarization of protons.

A. Derbenev, Skrinsky, and Kondratenko proposed a scheme with 5 transversely deflecting magnets in a straight section, deflecting alternately in the horizontal and vertical directions. Each magnet causes a 90° precession. The longitudinal polarization in the diametrically opposed straight section would be the same on every revolution.

B. This configuration is stable against small magnet perturbations, and the beam passes through no resonances of any kind during acceleration.

C. Two additional magnets are needed to restore the equilibrium orbit. The strengths of all 7 magnets are energy independent; 6 would have a strength of B·L = 27 KG-m each, and the 7th twice as strong.

D. An alternate configuration uses 3 magnets deflecting in the transverse direction at successive angles of 0°, 120° 240°. In this case the spin in the opposing straight section is radial, not longitudinal, on each revolution. This configuration also requires 2 additional magnets for equilibrium orbit restoration.

E. More work is needed on these schemes to determine their susceptibility to small deviations, and to study the feasibility of combining these with other, conventional systems.

4. Faster polarimetry than is presently available is almost a necessity for acceleration of polarized protons to higher energies.

EXPERIMENTAL SESSION

October 21 Afternoon

O. Chamberlain and M. Marshak presiding

1. We expect to need superior polarized targets, especially where inclusive reactions are involved, as the scattering from hydrogen in a conventional target cannot be separated kinematically from bound protons. This has led us to look again at HD polarized targets as recommended by Honig (at ANL meeting, June 10, 1977). If magnet fields of 10T and temperatures of 25mK are used to polarize by brute force methods, a plausible target polarization would start at 32% for protons (38% for full equilibrium). Practical experiments require polarization reversal, so it's important that Honig's comments on adiabatic fast passage be considered as a means for target reversal.

On the basis of these considerations, the HD target is hardly superior to conventional targets, so a switch to HD targets is not called for at this time.

2. Experiments with a Λ in the initial or final state provide considerable information about spin. Overseth is planning a Fermilab experiment on $\Lambda p \rightarrow \Lambda p$ with a polarized Λ beam and a self-analysis of the final state Λ by the decay. He will measure the Wolfenstein parameter R (also known as D_{SS}) which will provide a test of Yang's model of rotating nuclear matter as an explanation for the rise of the pp total cross-section.

3. The workshop group considered the kinds of polarization that would be of interest to experimentalists in an intersecting beam geometry, such as exists at the ISR. This geometry involves beams intersecting in a horizontal plane, and the useful polarizations states are:

a) nn (both spins normal to the plane)
b) $\ell\ell$ (both spins longitudinal)
c) ss (both spins sideways)
d) sℓ (one sideways, one longitudinal)

SOURCE WORKSHOP
October 21, Afternoon
H. Glavish presiding

1. Polarized H^- ion injection via stripping:

The scattering rms half-angle is approximated by

$$\varphi_s \simeq 0.25 \sqrt{\frac{Z_2(Z_2+1)}{A_2}} \cdot \frac{Z_1\sqrt{t}}{E_1} \text{ mrads}$$

where E_1 is the incident energy (MeV); Z_1 is the atomic number of the incident particles; Z_2 is the atomic number of the stripper; A_2 is the mass number of the stripper; and t is the stripper thickness in $\mu gm/cm^2$. The above φ_s applies to a single pass through the stripper; $\varphi = \varphi_s\sqrt{N}$ for N passes. For effective stripping $t \geq 0.3\ \mu gm/cm^2$ is best; the thinnest carbon foils which can be used are $\simeq 5\ \mu gm/cm^2$.

Conclusion: scattering is not a limitation at 50 MeV, even for 1000 turn injection.

Gas jet strippers are probably best in practice, since foils have a limited lifetime(e.g., a sodium vapor jet).

2. Effectiveness of polarized H^- over polarized H^+:

An effective 10 turn (for example 20 turn at 50% efficiency) injection at 50 MeV is a practical figure for H^+ ions.

An effective 100 turn injection at 50 MeV is probably OK for H^- ions, with stripping at injection. The limit is not scattering from the stripper, but the linac pulse time (600 μsec for 100 turn injection into CERN PS) and the difficulty of maintaining $\dot{B} \simeq 0$ stability over a long period. In a specially designed system it may be possible to achieve 200 turn injection.

Conclusion: Polarized H^- ions are 10 (perhaps 20 in special cases) times as effective as polarized H^+ ions for injection, in the regime where the space charge limit is not reached (i.e., for H^- ion currents less than a few mA).

3. Low energy collector ring - further considerations:

By designing a ring to operate at 750 keV the space charge limit can be increased to 8×10^{11} particles. However it does not seem feasible to utilize extra particles because of the linac acceptance. Furthermore, it does not appear feasible to inject 8×10^{11} particles into a collector ring as this would require 250 to 1000 turns (depending on the ring radius). Injection by H^- stripping at such a low energy would produce too much scattering.

4. Colliding beam polarized ion source of 1mA H^-:

The principles were discussed. We posed the practical problem of space charge neutralization and ion extraction.

EXPERIMENTAL SESSION

October 24, Morning

O. Chamberlain and M. Marshak presiding

1. How to see high $P_\perp$ jets:

   A. A. Yokosawa led a discussion of 3 FNAL experiments 236, 260, and 395.

   B. At HE a jet in the cm system appears as a jet in the lab system also. One can therefore look for jets in the lab frame (E395).

   C. The existing results are in conflict regarding the $P_\perp$ dependence of the invariant inclusive cross section.

   D. The same detection techniques in jet production appear to be important when polarized beams and targets are employed.

   E. According to the review talk of G. Thomas, we expect that 2-spin measurements (such as A_{nn}, $A_{\ell\ell}$) will show larger effects than 1-spin observations (the polarization).

2. Discussion of Λ beams (talk given by O. Overseth):

   A. Overseth reviewed Λ detection and its decay process, and Λ polarization measurements.

   B. The measurements of differential cross section for Λp and $P_\perp$ dependence of the Λ polarization were discussed. The Λ polarization appears to be large at P_Λ below 120 GeV/c but consistent with zero at P_Λ above 120 GeV/c, for $t \simeq -0.2$.

   C. At HE, processes with a final Λ will remain important to experimentalists because of the self analyzing Λ spin.

3. Discussion of elastic pp scattering (talk given by A. Krisch):

A. Use of the variable $\rho_{\perp}^2 = \beta^2 P_{\perp}^2 \sigma_{TOT}/38.3$ mb removes the energy dependence in the diffraction region and in the "diffractive" large $P_{\perp}^2$ region.

B. There are three regions with slopes (in the $\rho_{\perp}^2$ variable) of about 10, 3.4, and 1.6. The middle region goes away with increasing energy.

C. The normal spin-spin coefficient A_{nn} appears to be large in the third region; the reason for this is not clear.

SOURCES WORKSHOP

October 24, Morning

H.F. Glavish presiding

1. Discussion on colliding beam H^- source. Two possible schemes were considered

A. A long interaction region with positive ion trapping.

B. A short interaction region which should eliminate the need to space charge compensate.

2. Magnetic separation of a 500 eV H° beam. The conclusion was that this was not considered practical.

ACCELERATOR WORKSHOP
October 24, Morning
E. Courant presiding

1. Tried to estimate the strengths of resonances and their scaling laws in machines for which detailed calculations have not yet been made: the consensus was that the strength of the depolarizing resonances increases as $\simeq E^{\alpha}$, where $2 \leqslant \alpha \leqslant 3$, and multiplied by azimuthal form factors. These azimuthal form factors are large when the order of the resonance is equal to the major periodicity of the machine, and smaller for other multiples of minor periodicities. For example, at the $G\gamma = 76$ at ISABELLE, the form factor is large. Crude computations show that the scaling law is approximately correct.

2. Depolarization rate in the steady state, when the beam gradually changes because of scattering or cooling: Montague presented a formula for the corresponding case in electron storage rings where one has damping due to synchrotron oscillations. It appears plausible that in considering protons, one should replace the electron synchrotron oscillation damping rate by the rate of change of amplitude of vertical oscillations. As a result it seems that the rate of depolarization in the steady state is approximately equal to the rate of change of the vertical amplitude, multiplied by the form factor, which is normally on the order of unity, depending on how close one is to a resonance. If the form factor is small, the depolarization will not be important, but if it is large, then the rate of depolarization will be faster than the damping of the vertical amplitude.

The afternoon session of this group was combined with the polarimeter workshop, which is summarized by J.B. Roberts.

THEORY SESSION

October 24, Afternoon

F. Low and G. Thomas presiding

1. Thomas and Margolis looked at the problem of using the Primakoff effect as a polarimeter ($pA \rightarrow p\pi^\circ A$). This seems to work very well, and they concluded that one could even measure longitudinal polarizations with this method. Thomas will give a talk on this subject Oct. 25.

2. No other people have reported results, but there was a long discussion on inclusive Λ polarization, and the large Λ polarizations seen in such reactions are not surprising. Soffer will try to calculate numbers for inclusive Λ reactions. The group concluded that the large polarizations in elastic Λp (compared to pp) are surprising, and that they could not think of a mechanism to understand it. Overseth and Heller discussed their experiment.

3. Yokosawa gave a half hour summary on the experimental situation: what kinds of experiments and the kinds of accuracy that were possible. The group discussed polarized beams from Λ decay and jet experiments, and concluded that large $P_\perp$ experiments using such beams would be very difficult. On the other hand, cross section differences $\Delta\sigma$ and spin dependence of amplitudes in the diffraction region would be much easier, and Yokosawa concluded that such experiments were feasible. Even polarized deuteron-deuteron reactions would be quite difficult because of the intensity losses and high backgrounds.

POLARIMETER WORKSHOP
October 24, Afternoon
J.B. Roberts presiding

1. Measurement of polarization of accelerating protons at 30 and at 300 GeV was considered.

A. At 30 GeV, polarimetry in $pp \to pp$ at $-t \simeq 0.1$ with a double arm spectrometer, and in $pp \to p+x$ or $\pi+x$ with a single arm were considered feasible.

B. At 300 GeV, polarization in Coulomb scattering from an unpolarized jet was considered. A luminosity of $\simeq 10^{34}/cm^2 sec$ with 10^{12} circulating polarized deuterons was considered. It appeared that in the range $-t \simeq .001 - .005$ about 10 pulses would give a 10% measurement.

C. At 300 GeV, one could use a polarimeter based on $pp \to p+x$ or $\pi+x$, calibrated with a polarized jet at SPS or a PPT at FNAL.

D. The possibility of a polarized electron jet was raised, but no conclusions were reached.

2. In general one should extract immediately above and below a resonance with a front porch, and compare.

3. Measurement of polarization of accelerating deuterons at 30 and 300 GeV was also discussed.

A. At 30 GeV, one could use polarimeters based on $p(n)p \to pp$ with two arms, or on $dp \to \pi+X$ with a single arm, for example.

B. At 300 GeV, one could devise a polarimeter based on the inclusive reactions $dp \to \pi+X$, but this would have to be calibrated by measurements at the SPS or FNAL.

C. Atomic physics techniques based on NMR measurements do not seem practical.

4. Polarized deuteron sources: since the d has spin components $m_z = +1, 0, -1$ the deuteron beam will have a polarization

$$p_d = \frac{N_+ - N_-}{N_+ + N_- + N_o} \leqslant 2/3$$

This is because generally 1/3 of the deuterons will be in the $m_z = 0$ state. Thus one must be careful about making certain spin measurements.

5. Gas jets at the ISR were discussed.

A. The group considered H jets of two kinds: a pulsed high density gas jet, and a low pressure DC jet. Since the prolonged use of an H jet at the ISR would spoil the vacuum, these systems might be impractical on a long term basis. The polarimeter would consist of four double arm spectrometers, two for each beam.

B. A metal vapor (e.g., Li) jet was considered. This technique would employ four single arm spectrometers, two for each beam. The system could be initially calibrated against elastic scattering using an H jet, and could then serve as an absolute polarimeter.

COMBINED MORNING SESSIONS
October 25

1. G. Thomas gave a theoretical discussion on the reaction $pA \rightarrow p\pi^{\circ}A$ as an absolute polarimeter. The calculations were performed by Margolis and Thomas. He discussed the implications of

   a) current conservation
   b) high energy kinematics
   c) rates and backgrounds (still to be considered)

2. The various session chairmen gave short summaries of their respective workshops. E. Courant, J. Roberts, H. Glavish, O. Chamberlain, and F. Low each presented a quick review of the last two days, the proceedings of which we have already summarized elsewhere.

EXPERIMENTAL WORKSHOP

October 25, Afternoon

O. Chamberlain reporting

1. The group discussed the reaction $p + Be \to \Lambda + x$ and the available $P_\perp$ for doing reasonably feasible experiments. Overseth's group claimed 10 Λ/pulse with $P_\perp^2 \geqslant 2(\text{GeV/c})^2$. Jet production up to $P_\perp = 4$ GeV/c has been done with a beam no more intense than a projected polarized beam would have. This is also true of the Λ generated proton beam. One could do even better with a polarized deuteron beam.

2. The session listed the reactions that they would like to include as examples in their report, providing they give reasonably hopeful numbers, that is, they are experimentally possible. The examples are:

$p_\uparrow p \to \Lambda X$
$p_\uparrow\ Be \to \psi X$
$d_\uparrow\ Be \to \psi X$
$\Lambda p \to \Lambda p$
$p_\uparrow p \to \text{jets} + X$
$p_\uparrow p_\uparrow \to \text{jets} + X$
$p_\uparrow p \to \pi + X$ or $p + X$
$p_\uparrow p \to pp$ (small angle Coulomb interference)
$n_\uparrow p \to np$ (small angle Coulomb interference)

3. Elastic pp scattering at large $P_\perp$ shows a behavior very different from that in the diffraction region, and polarization effects there are likely to be large. There are probably also more interesting polarization effects to be seen at large $P_\perp$ than in the diffraction region.

4. The group reviewed the experimental uncertainties quoted in G. Fidecaro's proposal to measure A_{nn} and P (the polarizing power) at the SPS.

5. For most experimental purposes, polarized deuteron beams were about as good as polarized protons. There is difficulty in projecting into the future, so the group assumed that polarized deuteron sources in future would be about 500 μA, but did not assume any optimistic figures for electron cooling.

The workshop also looked at the question of highly energetic collisions between polarized deuterons ($d_\uparrow d_\uparrow \rightarrow$ particles). One finds that for d beams at FNAL, s = 380, whereas one would have s = 900 at the ISR. Here s is the center of mass energy squared between one of the target nucleons and one of the beam nucleons.

6. Larry Jones gave a talk in which he discussed polarized neutrons. His proposed ZGS neutron program is as follows:

a) verify the Schwinger effect
b) develop a polarimeter
c) determine the neutron beam polarization
d) measure $\Delta\sigma_L$ and $\Delta\sigma_T$ for np
e) measure ρ = ReA/ImA (spin averaged for np)
f) measure D^n_{NN} in np elastic, if there is enough intensity

He also proposed a FNAL neutron program:

a) make a polarimeter and a polarized beam of neutrons
b) measure $\Delta\sigma_L$ and $\Delta\sigma_T$ for np
c) measure A^n for np.

POLARIMETER WORKSHOP
October 26, Morning
J.B. Roberts reporting

The group looked at measured numbers for cross sections and determined the time span required to measure the beam polarization to within a given statistical error, for various choices of solid angle.

The group concluded that there are several types of feasible polarimeter for a Λ generated proton beam (10^6-10^7 pppp). For accelerating particles such as polarized deuterons, the choices are more restricted. An especially appealing polarimeter uses the asymmetry in π° production of an internal jet using similar apparatus to that planned for Dick, Cool, et al at SPS.

EXPERIMENTAL WORKSHOP

October 26, Morning

O. Chamberlain reporting

1. The group reviewed the luminosities obtainable at existing accelerators for $d_\uparrow d_\uparrow$, $d_\uparrow p_\uparrow$, $p_\uparrow p_\uparrow$, etc. They compared

   a) $d_\uparrow$ at SPS or FNAL colliding with a polarized H gas jet

   b) Λ decay generated $p_\uparrow$ colliding with a PPT

   c) $d_\uparrow$ on $d_\uparrow$ at the ISR

   d) $d_\uparrow$ on a PPT at FNAL

These are all experiments to measure A_{nn}.

2. Polarized neutron beams were discussed again. It seems that higher intensities than those proposed by L. Jones may be possible. It was not possible to predict ISABELLE luminosities because of lack of information.

3. Louis Michel gave a talk in which he discussed polarization as a four-vector, and the associated theoretical formalism.